Arzu Uraz

lture for Regenerating Cities

Arzu Uraz

Culture for Regenerating Cities

What can Istanbul 2010 learn from the European
Capitals of Culture Glasgow 1990 and Lille 2004?

LAP LAMBERT Academic Publishing

Impressum/Imprint (nur für Deutschland/only for Germany)
Bibliografische Information der Deutschen Nationalbibliothek: Die Deutsche Nationalbibliothek verzeichnet diese Publikation in der Deutschen Nationalbibliografie; detaillierte bibliografische Daten sind im Internet über http://dnb.d-nb.de abrufbar.
Alle in diesem Buch genannten Marken und Produktnamen unterliegen warenzeichen-, marken- oder patentrechtlichem Schutz bzw. sind Warenzeichen oder eingetragene Warenzeichen der jeweiligen Inhaber. Die Wiedergabe von Marken, Produktnamen, Gebrauchsnamen, Handelsnamen, Warenbezeichnungen u.s.w. in diesem Werk berechtigt auch ohne besondere Kennzeichnung nicht zu der Annahme, dass solche Namen im Sinne der Warenzeichen- und Markenschutzgesetzgebung als frei zu betrachten wären und daher von jedermann benutzt werden dürften.

Coverbild: www.ingimage.com

Verlag: LAP LAMBERT Academic Publishing GmbH & Co. KG
Heinrich-Böcking-Str. 6-8, 66121 Saarbrücken, Deutschland
Telefon +49 681 3720-310, Telefax +49 681 3720-3109
Email: info@lap-publishing.com

Herstellung in Deutschland (siehe letzte Seite)
ISBN: 978-3-8473-0671-9

Imprint (only for USA, GB)
Bibliographic information published by the Deutsche Nationalbibliothek: The Deutsche Nationalbibliothek lists this publication in the Deutsche Nationalbibliografie; detailed bibliographic data are available in the Internet at http://dnb.d-nb.de.
Any brand names and product names mentioned in this book are subject to trademark, brand or patent protection and are trademarks or registered trademarks of their respective holders. The use of brand names, product names, common names, trade names, product descriptions etc. even without a particular marking in this works is in no way to be construed to mean that such names may be regarded as unrestricted in respect of trademark and brand protection legislation and could thus be used by anyone.

Cover image: www.ingimage.com

Publisher: LAP LAMBERT Academic Publishing GmbH & Co. KG
Heinrich-Böcking-Str. 6-8, 66121 Saarbrücken, Germany
Phone +49 681 3720-310, Fax +49 681 3720-3109
Email: info@lap-publishing.com

Printed in the U.S.A.
Printed in the U.K. by (see last page)
ISBN: 978-3-8473-0671-9

ERASMUS UNIVERSITY OF ROTTERDAM

Thesis for a MA Degree in Arts & Cultural Studies,
Cultural Economics & Cultural Entrepreneurship

Culture for Regenerating Cities: What can Istanbul 2010 learn from the European Capitals of Culture Glasgow 1990 and Lille 2004?

Arzu Uraz

arzu.uraz@yahoo.com

Student Number
305839

Supervisor: Prof. Dr. Arjo Klamer
Second reader: Dr. Mariangela Lavanga

June 25, 2007
Rotterdam

Culture for Regenerating Cities:
What can Istanbul 2010 learn from the European Capitals of Culture Glasgow 1990 and Lille 2004?

By: Arzu Uraz

Abstract

In our contemporary era, the advocacy for culture has been articulated over many fields. Within the era of late capitalism culture is now heavily incorporated in regeneration and image (re-)making strategies of cities. The *European Capital of Culture (ECOC)* programme has become to serve as a catalyst for urban regeneration goals in cities. Among them, Istanbul is going to be an ECOC in 2010. By that time, Istanbul has a long way to learn from former ECOCs to understand the phenomenon and practice of culture-led strategies in urban planning and development. It is crucial for Istanbul to sustain the long-lasting impact for continuity.

The thesis is based on three pillars. First, it examines why cities pursue culture-led strategies in urban regeneration and how culture can contribute to urban regeneration in social and economic terms. Second, it finds out what academics, local authorities and Istanbul 2010 actors can learn from the highly debated cases of Glasgow 1990 and Lille 2004. Third, the theory and analysis of case studies of Glasgow and Lille offers projections for the Istanbul 2010 event and open space for future research on the challenges.

By performing a comparative analysis of case studies, the main finding is that culture has mainly contributed to city life in short-term. However, main problems in using the ECOC as a driver for urban regeneration emerged from high economic expectations and lack of clarity in long-term projections. Also, economic and social targets have fallen short in achieving sustainability and integrity. The remedy is to assure that culture-led urban regeneration is backed up by concrete measures directed to economic and social issues separately. It is also found out that culture is way to far to solve structural problems of the city by itself.

Table of contents

Acknowledgements

My one-year Master's experience in the Netherlands has been an invaluable life time experience for me. Although the thesis process in the end got quite hectic -as usual among all graduate students-, I believe the challenges; unforgettable friendships and in-depth academic conversations I have all gone through will lead my way.

From the very beginning till the end of this experience, there are many people I am indebted. First of all, I am grateful to my supervisor, Prof. Dr. Arjo Klamer and my co-reader Dr. Mariangela Lavanaga for their guidance, support and inspiration I have received through out the entire process. Secondly, I would like to express my deep gratitude to Dr. Filip Vermeylen for his generous support and his never-ending efforts to maintain our Master's programme. Besides, I am thankful to Ms. Paula van der Houwen and Ms. Sabaï Doodkorte, from the Faculty of History and Arts, who have tirelessly helped in my arrival formalities to the Netherlands.

I am also sincerely indebted to Assoc. Prof. Dr. Ayşen Savaş, Advisor to the President of Middle East Technical University and an Architect, and Assoc. Prof. Dr. Selim Çağatay, Hacettepe University, for their trust in my passion for cultural economics.

As it was a challenging procedure for me to be an international student in the Netherlands, my dear Dutch friends Carlien Schrijvershof, Froukje Budding, and Marianne van de Velde always made me feel at home. I enjoyed every moment of their friendship and cherished their opinions on my personal and academic development. Also, I am thankful to Elena Bird wholeheartedly, for her warm and supportive friendship and her valuable contributions to my thesis in every stage.

Last but not least, I send my warmest thanks to my following friends in Ankara for their endless support: Aslı, Ece, Gözde and Şebnem. Also, my gratitude to Levent Orsel and Özlem Yetkin is indescribable: Thank you for believing in me.

Above all, I am greatly indebted to my beloved parents and my one and only brother Emre, for their trust, understanding and full-hearted support that helped me to follow my dream.

1. INTRODUCTION

"Istanbul: The City of Four Elements"

This is how the story of Istanbul as a Capital of Culture begins. In 2000, Istanbul set a target not only to be a top city for arts and culture in the world but also increase its competitiveness through culture. Like other mega cities, Istanbul will use its forthcoming European Capital of Culture 2010 designation as a magic instrument for reanimating the cultural heritage and sites that the city has owned for centuries. However it goes further. While improving the economy was once based on economic tools today culture and its instruments exert an effect for economic and social outcomes. City-growth, through culture has become a powerful mean of city branding in the competition for talented workers and lucrative tourism.

The recent comprehension of culture has tremendously altered compared with what it was two decades ago. In the past, culture received little attention with only for its heritage activities which governmental funds were making the structure more cumbersome. In the book *Culture and Local Development* by OECD (2005), the perception for culture in the past was basically stated as an unproductive source of activity (2005:7). Apparently, culture hardly appeared in economic and social planning or in any other development agendas. In those times, the economic potential of culture was not even desirable to be assessed. More importantly, its impact was assumed to have less to offer than the other economic activities - like manufacturing- (OECD 2005).

During the late 1980s, when globalisation was more salient than ever the discourse on multi-ethnic cosmopolitan city and spatial planning became increasingly coined in regional identity issues (See Evans, 2001; OECD, 2005; Kunzmann, 2004). After then, sustainable development began to consider culture as an important segment to economic and social capital. Culture became the cement between economic and social structure. It played a fundamental role in enhancing the system of shared values and national identity through strengthening cultural rights. In return, it enhanced democracy and social integration which all increased social capital (See OECD, 2005; Evans 2001). Culture was not perceived solely for its contribution to social values. For example, the notion for Klamer (1996) had dual meanings: System of shared values (general) and arts (narrow). Both in academic literature and in policy making, a lot of contributions of culture started to be realised. Adding to its aesthetic value, the utilitarian functionality of cultural activities formed a new frontier (OECD, 2005:8). In this new frontier the export potential of cultural goods increased as the economy got more global. In cities, where cultural amenities were rising as culture was getting more centred in city and spatial planning, culture also added to the entrepreneurial and innovation processes of cities. This was mainly noted as the Creative Milieu by Törnqvist (1983). Culture also changes the creative processes within cities or regions (Kunzmann, 2004). It promotes creative action in city development (Landry, 2000) whereas mutually interacts with the creative class that can also change the culture of a city (Florida, 2002).

As Kunzmann (2004) states the old paradigms of social and sustainable development loosened their former appeal and culture became more prominent in both policy-making and planning. For Evans (2001), globalisation and late capitalism has formed an ideal paradigm for incorporating culture in urban regeneration and development strategies. As interaction between the symbolic economy and political economies of culture has increased, "Urban Renaissance" has emerged. In this new paradigm, urban renaissance incorporated culture as a consumption, production and image strategy in cities-regions in developed, and even in developing and emerging towns (Evans 2001:2).

Among the various forms of instrumentalising culture, culture-led urban regeneration projects have typically flourished in West European countries, yet this trend is changing. Today emerging cities in developing countries have also set out initiatives for utilizing their cultural potential, both in economic and social terms. Culture-led urban regeneration has broadened its territories to embrace the "new" growing mega city. Whereas before it was favourable to apply this approach to "old" or "declining" cities, now the phenomenon has reached growing mega cities in Japan, Thailand, South Korea, China and even the Arab world. Following the UNESCO's Global Alliance for Cultural Diversity initiative, cultural industries in developing countries are sought to encourage knowledge-sharing, capacity building, good practice and mentoring between members[1]. Furthermore, Shanghai, Dubai and Sao Paolo are becoming more attractive for business investors in terms

---

[1] http://portal.unesco.org/culture/en/ev.php-
URL_ID=24478&URL_DO=DO_TOPIC&URL_SECTION=201.html

of boosting creativity via cultural industries. Moreover apart from the European continent, Cultural Capital programmes have also been designed in the Arab world and in Latin America.

As the trend becomes more complex and its practice spreads out beyond the territories of the developed countries, it is crucial for the local authorities and the decision makers in the cultural terrain to analyse the role of culture in general and then, magnify the lens towards urban regeneration. The growing complex nature of culture has called for more interdisciplinary approaches and frameworks to understand the phenomenon and practice of culture-led strategies in urban planning and development. Recently there are many researches undergoing to understand the phenomenon from different glasses. This thesis chooses to use the glasses of cultural economics. It aims to make the reader acquainted with the notions, aspects and critiques of cultural economics on the increased role of culture in sustainable development in general and urban regeneration in particular. Up till now, it is explicitly shown how culture came to its heydays today. However, this also brings up many questions into our sceptic mind that if all these developments, cultural programmes really work out or not. Or are they just illusions which are waste of money? And finally, what do recent studies suggest about reality?

All the above mentioned questions and more are the scope of this thesis to answer through the critical look of cultural economics. The thesis at hand aims to answer the initial question of why cities are pursuing culture-led strategies in urban regeneration and how culture can contribute to urban regeneration in social and economic terms in the context of cultural economics discussion. Secondly, its aim is to find out what academics, local authorities and Istanbul 2010 actors can learn from the highly debated cases of Glasgow 1990 and Lille 2004, which were both *European Capitals of Culture*. The theory and analysis of case studies of Glasgow and Lille will offer projections for the Istanbul 2010 event and open space for future research on the challenges of this kind of approach. The previous work done on these two cases also shows what kind of definitions, approaches, sources and tools were used in evaluating the outcomes. Since there are very few studies that have focused on Istanbul 2010 especially with the emphasis on the case study analysis, the present study basically has the main task in raising the debate around Istanbul's future experience in cultural development.

The motives behind choosing Glasgow and Lille as case studies are:

i. Size matters: Since Istanbul will be the largest ECOC; the case studies should be selected carefully. The cultural programme implementation requires a well-defined job description and an effective and efficient management. It is known that the bigger the city gets the more complex the structure is. Even it is very hard to find a city at the same scale as Istanbul, in which the total resident population is nearly 10,020,000, Glasgow is the biggest city in Scotland with a resident population of 1,749,154 and Lille is the forth largest city in France having 1,143,125 people in 2003[2].

ii. Both Glasgow and Lille are renowned cases for applying culture-led urban regeneration field. Glasgow has been the first city to use the ECOC event for urban regeneration purposes (Lavanga, 2006: 5). Many studies have been done, some have promoted their strategies and some have harshly criticised them. Whilst, Lille has a relatively more accepted status in terms of social inclusion than Glasgow has.

iii. Both cities were once mostly referred as highly industrialised cities with their bulky manufacturing sectors and later, declining cities for their post-industrial pattern. Today, together with Istanbul, all these cities are characterised by the post-Fordist economic trend. Besides, Glasgow and Lille had suffered from extensive suburbia. Similarly, Istanbul has been struggling with the socially deprived community that was resulted by the dual economic structure.

One of the strongest contributions of the study will be the analysis of how culture-led urban regeneration can help a city like Istanbul to make better use of its cultural amenities in urban development through learning from past examples which have sought the similar projection. The other point will be to see how the ECOC programme will be helpful to establish long lasting-impact for Istanbul in order to reach a sustainable city.

The challenges to be discussed mainly come out from the city's specialities. By the year 2010, Istanbul will be the largest ECOC, with a population of nearly eleven million people. Moreover, Istanbul 2010 agenda embraces a multi-faceted approach in its targets and looking forward to building longer-term impact of this event

---

[2] See Eurostat Online Database, Urban Audit: http://epp.eurostat.ec.europa.eu/

together with planning culture-led urban regeneration. Indeed, with the ECOC experience the city will have much to offer and demonstrate distinct trends to be analysed in the future.

For centuries, Istanbul has served as a centre of industrial activities and rapidly growing as the financial hub of Turkey. Beyond its economic features, Istanbul has a long and rich history of culture. Many decision makers and the cultural planners of Istanbul 2010 advocate that the city can take the advantage of its heritage through refurbishing it and adding new cultural clusters to increase its potential in cultural tourism.

Another speciality of Istanbul is that the city has room to grow and has not yet suffered the syndromes of declining cities. After the ECOC event takes place, this would suggest interesting implications for the literature to revise the widely acknowledged perception that culture-led strategies in depressed cities work more than in the other un-depressed ones.

In terms of organisation and managerial issues, Istanbul might prove to be an interesting model to analyse. As it is stressed in the application book of Istanbul's ECOC initiative, *Istanbul: City of Four Elements*[3], the organisational structure and the network of partnerships are expected to positively affect the public-private partnerships in the future. In one way, building social networks and continuing them are important for sustainability of the cultural sector to receive long-lasting effects. The other is that ECOC can also stimulate cultural entrepreneurship in this process. The cultural entrepreneurs preparing for Istanbul's 2010 event will take the advantage of building innovative and creative partnerships in the cultural and creative sector. In return, this will strengthen the cultural connections in the web of networks.

Moreover, the ECOC experience will bring many novelties to Istanbul. It will strengthen Istanbul's urban governance by improving cultural citizenship. New megapolis administration models will be built. Above all, all these novelties will be insightful for other cities similar in scale and social structure with Istanbul.

---

[3] The *Istanbul: City of four elements, Application Book* can be found online at:
http://www.istanbul2010.org/?p=103&lang=tur

The reasons for a comparative analysis are three-fold. First, it facilitates in deriving both the positive and negative aspects of the culture-led urban regeneration phenomenon. Second, it adds for a better understanding with clear implications for future discussion about the ECOC programme and Istanbul's future urban development goals in particular. Especially, in the case of Istanbul the comparative analysis guides the study why the city should consider ECOC as a tool for triggering culture-led urban regeneration. Least but not last, the general conclusions drawn from the economic and social outcomes of the case studies in this thesis are neither to imply that same projections of ECOC should be implemented to achieve a culturally regenerated city nor to perceive the issue as a one-size-fits-all project, but rather to bring good practices with respect to their balanced combination of economic and social aspects to the discussion floor.

1.1 Terms and Definitions

Since it has always been a problem to draw the borders of culture, here in this study the notion will comply with its two main senses: The narrower sense, where cultural goods and services are meant together with cultural industries and the wider sense in which ethnic components, traditions and its social connotations are defined in a system of shared values. The reason is that it would have been too much of restriction to the economics if culture was only meant to be used for its narrower definition. It is thought that culture in urban regeneration strategies and programmes have an important social role in addition to recognising the economic influence it has. However, when it comes to concrete or measurable outcomes for indicating the ECOC process, the term "culture" is mostly referred to "cultural industries", where the cultural production and consumption of the cultural goods and services are created.

If this thesis had been written twenty-years ago, the definition would have been as what traditionally was understood as art and popular culture (Wynne, 1992:1). However, as Wynne puts forward very bluntly, the increasing "cross-overs" between different forms of activities has broadened the definition. Together with the rise of the knowledge economy, the term "cultural industries" is interchangeably used for "creative industries" or for some authors, "creative industries" are more than the traditional cultural industries plus the creative business services (new media, advertising, fashion, design, architecture, Creative

ICT and software). Especially, the latter category is included for its high degree of intellectual property embedded in content. In other words, 'cultural industries' are also known as 'copyright industries' (Throsby, 2001:112).

Since the literature includes many studies on cultural industries recently, there hasn't been a clear consensus on the definition yet. The definition from the Department of Culture, Media and Sport in the UK is frequently used[4]; however, in this present study it is appropriate to use the definition from Eurostat, which serves better for constructing the two case studies-Glasgow and Lille- on a commonly used definition.

The Eurostat-LEG definition of culture[5]:

Artistic and Monumental Heritage - Historical monuments - Museums - Archaeological sites - Other heritage	Visual Arts - Visual arts (incl. design) - Photography - Multidisciplinary
Performing Arts - Music - Dance - Music Theatre - Drama Theatre - Multidisciplinary - Other (circus, pantomime, etc.)	Book and Press - Books - Newspapers and periodicals
Archives	Audio and Audiovisual Media/Multimedia - Film - Radio - Television - Video - Audio records - Multimedia
Libraries	
Architecture	

---

[4] The definition is based on seven domains including: Audio-visual (film, television, radio, new media, music); Books and Press; Heritage (museums, libraries, archives and historic environment); Performance (theatre, arts and dance); Sport; Tourism (including gambling and betting activities); Visual Arts (galleries, architecture, design and crafts)
http://www.culture.gov.uk/www.culture.gov.uk/Templates/Publishing/Research.aspx?NRMODE=Published
&NRNODEGUID=%7bC0E53EED-B0BD-41ED-BC1D-
B717723781A1%7d&NRORIGINALURL=%2fReference_library%2fResearch%2fdet%2fglossary_abbrevi
ations%2ehtm&NRCACHEHINT=NoModifyGuest#top
[5] European Commission, Cultural Statistics in the EU: Final Report of the LEG, 2000, Luxembourg, pg. 24-35

1.2 Methodology

The research at hand pursues a comparative methodology for a two-sided approach instead focusing from a biased perspective by examining only a single successful case. When the research questions are taken into account, approaching to the research topic by providing both the positive and the negative aspects of the culture-led urban regeneration issue of the specific two cases is crucial for the sake of the research.

The cases, hereby Glasgow and Lille, are analysed in depth according to the following categories:
1. The objectives and targets of the programmes are briefly set out
2. The implementation:
- How the goals were met
- Types of collaborations
- Specific mechanisms, organisational structure designed
3. Outcomes
- Economic
- Social
- Cultural,
- Other, if available

The problematic issues emerging from both cases will be examined under a separate chapter divided into the following sub-chapters:
4. Problems
- Economic Problems
- Social and Cultural Problems
- Other (including organisational, governmental and etc.)

1.3 What Can Cultural Economics Contribute?

The question stated as the heading of this sub-chapter dares fruitful implications in understanding the role of culture in urban development plans from the viewpoint of cultural economics.

In this thesis cultural economics enlightens several concepts. First, it helps us understand the economic perspective of the role of culture as a catalyst in urban development. Especially, in the case-study analyses of Glasgow and Lille, cultural economics would help clarifying the organisation and the economic features of the ECOC programme.

Second, cultural goods and services provided by cultural industries have distinct characteristics compared to other economic goods. Since they are mostly attributed as experience goods, different and interdisciplinary approaches need to be applied for coherent analysis. At this point, cultural economics well serves in understanding the contribution of cultural industries to economic development. It brings out the question if those industries are different from other industries, and if so what are their peculiarities? Also, it stirs the discussion in being sceptical while the outcomes and problematic issues of the case studies are discussed in chapter 4. It is discovered that certain limitations appear while assessing the impact of culture on cities. Even though cultural economics enlightens the organisational and economic features in particular, it does not restrict its approach merely on analysing the economic impact of culture. It stresses the pitfalls of looking only from one side to the subject matter. As cultural economics stresses the peculiarity of cultural goods, it brings up interdisciplinary viewpoints into the discussion as well.

Third, what cultural economics can bring into the research is providing a new way of looking at problems such as cultural pessimism. The recent criticism of using culture as a main instrument of urban development issues evokes the discussion of cultural pessimism that we find with several critics. What these critics highlight is that culture is under the threat of being homogenised worldwide by global capitalism and hence the fierce race of transforming cities into new cultural hubs hinders the artistic innovation and reduces cultural diversity (see Evans, 2001). But not everyone agrees. For example Tyler Cowen opposes this statement in his book *In Praise of Commercial Culture* (1998), arguing that globalisation, instead, can diversify cultural production and stimulate new cultural forms in the market.

Fourth, cultural economics can lead the analysis by differentiating the notions of cultural capital into two forms: intangible and tangible. The intangible - cultural networks, pride, cultural rights, governance - and the tangible - theatres, galleries, museums, places of cultural production. So as strategies and projects are developed for urban regeneration both tangible and intangible forms of capitals can be identified and be subjected to long-term impact and assessments.

As the debate continues around the cultural or creative city projections, Throsby (2001:124) has stressed how arts have become more important in economic life through "embracing the broader issues of the urban fabric".

An argument can be provided by Throsby's statements to the question 'why culture is so dear to cultural economics?' The reason is explained in Throsby's own words:

> "...neglect of cultural capital by allowing heritage to deteriorate, by failing to sustain the cultural values that provide people with a sense of identity, and by not undertaking the investment needed to maintain the stock of tangible and intangible cultural or to increase it (for example by the production of new artworks) will likewise place cultural systems in jeopardy and may cause them to break down, with consequent loss of welfare and economic output" (Throsby in Towse (ed) 2001:183)

In the discourse of culture and its contributions to regional growth, Throsby (2001) states four roles for culture in the life of cities.

1. A specific cultural amenity which serves as a catalyst
2. The notion 'cultural district' or 'cultural quarter' may facilitate the development in the local area.
3. The cultural industries may act a vital component for city's economy
4. In the wider sense of culture, the role can add to community identity, creativity, social cohesion among city inhabitants

The economic outcome of these roles can be assessed with the help of cultural economics. The first impact results from the direct revenues of cultural activities which in turn contribute to the local economy. Second, the indirect effects can be observed in the revenue increase of other supplementary businesses, such as restaurants, hotel and transport services etc. Third, according to Throsby, the expansion in employment opportunities brought by those cultural amenities can be significant. Although it is out of the scope of this paper, the employment creation ought to be analysed in depth to see if the pattern of job types is more aligned with stable or rather being seasonal or temporary jobs. Fourth, Throsby also proposes that in the post-Fordist period culture might help city economies depressed from industrial decline. In addition to all these concrete economic impacts, there are externalities in other fields of the economy that can improve a city's image and local pride.

It is thought that especially in the current period where one searches for a holistic model to comprehend globalisation and its impacts on both economic and cultural life, the interdisciplinary approach of cultural economics seem appropriate to work with in order to understand the complex interaction between the economic, social, cultural and environmental objectives of urban development. The discourse of urban development has recently given rise to the notion of 'sustainable city'. Throsby (2001) has grounded the discourse on a "sustainable city model" which focuses the attention on environmental concerns by working on measures such as improved transportation, efficient energy usage, re-cycling and waste management and together recognising the vital cultural values of national pride, identity, creativity which all could be enhanced through setting a culture-led urban regeneration agenda.

Above all, cultural economics contributes to the thesis at hand through giving insights on culture and development, economic role of culture, organisation of cultural industries and questioning the cultural indicators compiled for impact studies.

2. THE RISE OF CULTURE: NOT THE 'CHERRY ON THE CAKE' ANYMORE

"Instead of culture springing from the inner workings of our cities, we see it as a way to make our cities work..." (Jonathan Glancey, The Guardian, 29 March 2003 cited in Wilks-Heeg, S. & North, P. 2004*).*

The quotation above precisely describes the current perception of culture today. Culture is now the new instrument for local economic development. A recent study on the *Economy of Culture in Europe* (2006) commissioned by the European Commission have published some figures about the cultural sector:

- The sector turned over more than 654 billions Euros in 2003
- The sector contributed to 2.6% of EU GDP in 2003
- The overall growth of the sector's value added was 19.7% in 1999-2003.
- In 2004, at least 5.8 million people worked in the sector, equivalent to 3.1% of total employed population in Europe.

Posing the "why" question at this point would be a good start to understand why culture is now so precious in terms of economy and social welfare.

Indeed, the alteration of the role of culture within our lives has lots to do with the overwhelmingly changing global economy. Initially, the "new economy" has brought the "knowledge" as the high-value added factor into the economy. Formerly, the economy was mainly driven by manufacturing which utilized mass physical capital. As a result, Fordist type of production became the main source of economic growth. Today, intellectual capital has stirred the knowledge economy and given rise to post-Fordist cultural economies in advanced capital societies[6]. Innovation with creativity now has a higher impact on economic growth (See Kunzmann, 2004; Florida, 2002). It is the era of intangibles, where symbolic meaning is emphasized and thus, everything is built on the experience economy (Pine and Gilmore, 1999).

Culture perfectly suits the new experiences for society because it embeds symbolic values. This happens best in cities. As Scott (2000) puts forward, culture helps communities differentiate themselves from one another because it is unique and peculiar to each city, even to each district. The "new symbolic economy" (Zukin,

---

[6] See Crane (1992), Lash and Urry (1994) in Scott (2000)

1995:1-2) fuelled by culture facilitates cities in improving their new cosmopolitan image (Hitters 2005: 5).

Another factor that increases the importance of culture and makes it more instrumental in policy and planning documents is the increased competition among cities. According to Scott (2006), competition is due to the factors that cities have dense networks in large agglomerations and globalisation can enhance the possibilities of vertical disintegration[7], productive agglomeration and specialisation (Scott, 1988 in Scott, 2000:5). He continues by asserting that cities are both complements in the sense that they are in exchanges of products and at the same time rivals to one another where they compete for securing inward investment, widening external markets and attracting visitors. Especially in the current era, policy makers have recognised the potential of cultural promotion tools in combination with local economic development programs (Scott 2006:10). Local authorities have become more flexible in forming their own programming for culture. This situation also allows them to cross-over different borders and bring new policy agendas for linking culture with the economy. Given the fact that agglomerations in cultural industries function well in the economies of cities (Hitters 2005, Hitters and Richards 2002, p.236 and Miles 1997), policy makers are now tailoring their documents accordingly to create new linkages between culture and other sectors.

Everything about the new ways of linking culture and the economy is due to the aim of regenerating city life in economic and social terms. Throsby (2001) has brought up the idea that arts and culture may have a more pervasive role in urban regeneration through fostering community identity, creativity, cohesion and vitality via the cultural characteristics and practices, which define the city and its citizens. This statement has justified the role of culture as *sine-qua-non* almost in all kinds of policy documents. Cities seeking to implement regeneration plans set up a story in which the cultural dimension is the core and let people *experience* that unique place.

---

[7] It refers to an organisational form of industrial production. In this form, production occurs in separate companies which each specialize on specific tasks. It is important for creative and cultural industries in sharing the risk. A good example for this is Hollywood. For more information see Storper and Christopherson, 1987.

As this trend becomes more and more popular, the outcomes of those stories are coming under more criticism. The policy documents, especially for creating competitive cities through fostering arts and culture look more the same as the world is globalising. The alikeness does not stem from what the targets or features are but from the rhetoric. As many scholars dislike this situation, they mostly argue that the targets are overestimated, social problems appear and they are disregarded, the impact studies are a kind of justification for the economic outcomes more than it is for the quality of life, and that mostly "north" cases are used for studies[8].

Keeping these disputable points in mind, the applicability of culture in local/regional economic development policies is not losing momentum. In fact it is growing to embrace a wider definition of culture. It is reaching out to both senses which culture is now perceived as its industries - cultural and creative- and its whole system of shared values that enriches diversity in societies it belongs to. Because of the influx of immigration on cities Charles Landry (2000) has asserted the necessity of multiculturalism as a prerequisite for fostering the creative industries in cities and thus, urban regeneration. Landry's assertion has also served for eradicating social problems through culture and re-justifying the importance of creative industries-driven regeneration. Montgomery (2003) notes a check-list for cultural quarter making and suggests that almost every city could adopt the model of culture-led urban regeneration by highlighting the economic potential of cultural clusters (in Newman, 2005).

It is interesting to see the alteration of the definition of culture in the discussion of culture-led urban regeneration. As the counter arguments rise upon the role of culture in cities, definitions or the content of the notion is altered to create a reliable or justifiable situation. As a result, saying "creative industries" rather than cultural industries and each time adding new sectors-which is always debatable-reflects the search for justifying policies in this field. Yet still, it would be better to be sceptical while dealing with these new definitions and its extensions.

---

[8] See Miles (2005), Miles and Paddison (2005), Isar and Bonet (2007)

2.1 The Rise of Culture in Urban Regeneration

Cities have been reshaping their image constantly. Ever since globalisation has started to shift the economy towards a more information-based, and subsequently to a creativity based economy in the 1980s, the growing potential of cultural industries drove the economy. As Bianchini (1993) states, especially declining cities have been using arts and cultural industries increasingly as means of city marketing. Consequently, as it strengthened the image and developed the cultural industries of the city culture-led strategies became more centred in cities' regeneration plans.

The 21st century became the heydays of culture-led urban regeneration, but why? Why were all those fancy "cultural hubs, industries, creative clusters" words used?

During the transition from industrialization to deindustrialisation after 1960s, some cities started to decline. Loss in employment opportunities resulted as out-migration from cities. The threat for cities that were experiencing lower economic growth and loss in attractiveness forced the cities to find ways out. This stimulated the cities to become more competitive in the global arena. As a way out, cities were looking for a sort of revitalization and this, what brought the urban regeneration. And culture became the key strategy for urban areas.

Strategies for renovating and investing in culture and creative industries have brought a new phenomenon into urban and cultural policy: The Culture-Led Urban Regeneration for cities. It is mainly built on cultural clusters and/or renovating the cultural infrastructure with the aid of modern technology to increase the cultural tourism potential of a city. As a turning point in this field, Glasgow received the greatest attention as being a model for culture-led urban regeneration[9]. Since the city was suffering from 'declining city' syndromes[10], the remedy was to bring the labour force back to work by remaking of the city. The remaking of the city was also related to its image. As a result, urban regeneration was needed to achieve this goal. The need of enhancing city image increased the rivalry among cities as well. And the key strategy to overcome this rivalry was hidden under the means of differentiation.

---

[9] See Garcia (2004b)
[10] See Gomez (1998a)

Urban regeneration gained substantial interest in today's post-industrialist or "post-Fordist" era. Cultural industries were entitled as the forces of cities' social and economic development. In particular, cities which sought to revive their image were going through a process in which cultural image was enhanced by the means of city branding. Furthermore, repeatedly cited in the report (2006) of "The Economy of Culture in Europe" prepared for the European Commission, cultural sectors have direct and indirect effects on social and economic development as well as on innovation and social cohesion.

Putting it shortly, the direct and indirect effects of cultural industries guides us to answer the question why they were used for urban regeneration.

2.2 The Factors Account for Culture in Regeneration

Many studies in the literature have tried to understand the usage of culture for cities' regeneration plans and to outline their importance in enhancing the economic life. Some studies strived to understand the dynamics behind this rising phenomenon and some have undertaken an evaluation of particular cases. Particularly, studies from the United Kingdom have immensely developed the literature in this field. Their main contribution was in the impact studies of culture in social and economic context (Evans & Shaw 2004, Myerscough 1988, Garcia 2004a-b, Evans 2005, Prentice & Andersen 2003, Gratton & Taylor 1995, Kelly and Kelly 2000).

The dynamics that highlighted the role of culture in urban regeneration was manifold. The factors stated in the literature prove to be diverse and complicated. The economic and social factors together with the changing political paradigm in the terrain of culture have led to a shift in trends in urban life. As Miles and Paddison (2005) put forward, the dynamics that have brought the "new orthodoxy" led to increase in citations, in policy documents and in urban regeneration.

Taking the social aspect of the issue, a report from UN's Habitat series (UNCHS, 2004) has underpinned the social rationale of culture. It emphasizes the opportunities that culture can generate in urban life especially in today's era where multiculturalism (cultural diversity) and exclusion concerns have become more prominent than ever. Today, like in many policy documents and in many cultural intiatives – ECOC -, the contribution of culture to social cohesion has also been highlighted.

Another factor which has catalyzed the process was the paradigmatic shift in cultural policies and the diminishing role of public intervention. As Miles & Paddison (2005) assert, the ideological de-legitimisation of state intervention and public-sector arts and media have changed the political background and thus opened more room for other private bodies to be players in the culture-led urban regeneration process. As the role of culture increased at local levels, the cultural policy arena shifted from cultural planning to cultural programming (Hitters, 2007).

Evans & Shaw (2004) have summarised the justification on the importance of culture in regenerating the urban economic and social life as follows:

The Economic Perspective:
- Inward investment (public-private sector leverage)   - Higher resident and visitor spend   - Job creation (direct, indirect, induced)/wealth creation   - Employer location/retention   - Retention of graduates in the area (inc. artists/*creatives*)   - A more diverse work force (skills, profile)   - A driver in the development of new business, retail and leisure areas   - More public-private-voluntary sector partnerships   - More corporate involvement in the local cultural sector (leading to support in cash and in kind)   - Increased property prices (residential and business)
The Social Perspective:

- A change in residents' perceptions of the place where they live
- Greater individual confidence and aspiration
- A clearer expression of individual and shared ideas and needs
- An increase in volunteering
- Increased organisational capacity at local level
- Increased social capital – 'the norms and networks that enable collective action' (World Bank)
- A change in the image or reputation of a place or group of people
- Stronger public-private-voluntary sector partnerships
- Reduced school truancy/offending behaviour
- Higher educational attainment
- New approaches to evaluation, consultation and representation
Source: Evans & Shaw (2004: 21, 29)

In fact, the contribution of culture and creative clusters is not only limited in economic and social terms. It also has a cultural, educational and environmental impact on the community. Local artists in the art market, individuals' taste formation and the infrastructure of the city are all affected by the process of culture-led regeneration.

Up till now, the factors behind culture in urban regeneration have clarified why culture meant so much for regeneration issues. The importance of culture as a driving force of regeneration has drawn large attention among the policy makers and local authorities. As a result, many policy documents focused on the role of culture and its contributions to urban regeneration. Nonetheless, there are studies which open room for discussing the advocacy behind the regeneration plans. As it is for the case in culture-led urban regeneration, bolstering culture and creative industries as the "great" driving forces can be somewhat misleading at the end when all the realities fall short from the expectations.

Some studies are introducing examples of good practices which have received great interest in global arena and have been used as models for other cities. Whereas some are pointing out the bad practices which have been too ambitious in their projections and have failed in reaching their targets in terms of audience numbers, profiles and income generated (Evans and Shaw 2004:7). Moreover, there are studies which are highlighting the failure from a more social perspective.

Many studies attempted to take a closer look to see whether the impact was really coming out or not. Some were arguing that the culture-led urban regeneration phenomenon was all a bunch of gloomy words for 'promoting' and nothing further than the political rhetoric (See Miles & Paddison, 2005; Miles, 2005; Evans 2005).

Like Miles and Paddison (2005) pose, the main concerns lead us to think whether we really do understand the complex nature of cultural investment and to what extent culture-led urban regeneration is more about rhetoric than it is about reality?

Social cleansing[11] or social exclusion as a result of gentrification[12] has proved to be of primary concern by many scholars in this field. "How and what you build through cultural industries?" are key questions for analysing the social effects. A culture-led urban regeneration merely consisted of architectural renovations and setting up new cultural industries or promoting flagship-events might yield detrimental effects in terms of social justice[13].

The complex nature of cultural investment also causes difficulties in measuring impacts and in having well-represented results. Most of the impact studies have tried to measure the role of culture in urban regeneration by relying more on indicators as number of visitors or spending through surveys. Evans (2005) states that "the emphasis on economic impacts produces headline-grabbing data about the raw potential of cultural investment, but it says next to nothing about the long-term sustainability of culture-led regeneration."

However...

Notwithstanding the reality of some bad practices in culture-led urban regeneration and its unanticipated detrimental effects, the role of cultural industries in regeneration hasn't shattered yet. Since this phenomenon became "popular", new projects and initiatives are increasing for regenerating cities' social and economic life via cultural industries and thus, reinforcing their image.

---

[11] See Gibson and Stevenson (2004)
[12] See Wilks-Heeg and North (2004)
[13] For discussion of culture and social justice see Stevenson (2004)

Today, not only in the post-industrialist but also in the developing countries cultural and/or creative industries are becoming more prominent in policy agendas. Many cities have developed cultural programmes, and most of them are grounded on regeneration activities. The concept of "Capital of Culture" has been widely adapted to nearly all of the continents in the world. Those are the *American, Arabian, Brazilian, Canadian, Catalan, Islamic, Volga and European Capital of Culture* programmes[14].

2.3 The *European Capital of Culture* and Istanbul

The *European Capital of Culture*[15] *(ECOC)* initiative was launched by the European Commission first as the *European City of Culture* programme in 1985. Later on, in 2004, it converted to its current name. The programme was initially sought to bring the citizens of the European Union together but afterwards the social and the economic effects made the initiative more far reaching than it was expected. As the trends and the paradigms altered in the cultural terrain, cities awarded with this title started to pave their own ways in this experience and each responded what culture meant for them in diverse forms in accordance with their distinct characteristics. The diversity was what made the programme each time a unique experience and in turn, produced various new models for the up-comers. With previous attempts in urban regeneration, Glasgow, becoming the ECOC in 1990, also proved to be a good practice in this initiative. It served as a model for many new comers[16].

Istanbul has different characteristics than other cities which have used ECOC as a tool for culture-led urban regeneration. It is a historically dominated city with a polycentric structure and shows no trace of a declining city even it has the post-Fordist pattern. The specialities Istanbul possesses will bring on interesting insights and challenges both to the city and the ECOC programme in many ways. By the year 2010, Istanbul will be the largest ECOC since ever this programme initiated, with a population of nearly eleven million people. Moreover, Istanbul 2010 agenda embraces a multi-faceted approach in its targets and looking forward to building longer-term impact of this event together with planning culture-led urban regeneration. Istanbul and the ECOC programme will mutually learn how culture-

---

[14] See the website for a detailed description of these projects: http://capital.culture.info/
[15] First, it was launched as the European Cities of Culture in 1985.
[16] See Garcia (2004) and (2005), Myerscough, (1988)

led urban regeneration can help a city like Istanbul to make use more of its cultural amenities in urban development through learning from past examples which have sought the similar projection.

Istanbul has competitive advantages. Beyond its economic characteristics, Istanbul's long and rich cultural legacy has contributed to its development in cultural amenities and thus, has already put Istanbul in an advantageous position in which facilitating the combination with local economic development programmes (Scott, 2006:10). As it follows, Istanbul is a historically dominant city and thus, has large accumulation of cultural capital and iconic buildings (Newman, 2005). And interestingly, as Newman (2005) asserts, historically dominant cities like Istanbul are less vulnerable to the changing trends in cultural development. He follows on as arguing that they can get some things wrong due to the fact of having a strong middle class and specialised jobs of the 'new economy'.

The city can take the advantage of its heritage through refurbishing it and adding new cultural clusters to increase more of its potential in cultural tourism and in other cultural industries. The other interesting point will be to see how the ECOC programme will be helpful to establish long lasting-impact for the city within this context. Yet still, unlike other European cities, Istanbul is evolving from a metropolis to a megapolis. It has more public space and has still room to grow. Moreover, it has not severely encountered the syndromes of declining or other post-industrialist cities yet.

Moreover, the ECOC experience will bring many novelties to Istanbul. It will strengthen Istanbul's urban governance by improving cultural citizenship. New megapolis administration models will be built. Above all, all these novelties will be insightful for other cities similar in scale and social structure with Istanbul.

In the future, Istanbul might prove to be an interesting model to analyse the effects of during and after the event. As it is also stressed in Istanbul's ECOC agenda[17], the organisational structure and the network of partnerships will positively affect the public-private partnerships in the future. However, the decision makers need to bear in mind the continuity of this structure so that sustainable partnerships will have fruitful effects. Apart from this, it would be

---

[17] See the bid document. Can be found online: http://www.istanbul2010.org/?p=103&lang=tur

insightful for its administrative model, urban governance, and for large scale emerging cities to see how the financial issues are tackled. It is very well-known that in the cultural terrain, building social networks and sustaining them are related to the relations of the third sector[18]. Being an ECOC will also stimulate cultural entrepreneurship in this process. The cultural entrepreneurs preparing for Istanbul's 2010 event will take the advantage of building innovative and creative partnerships in the cultural and creative sector.

---

[18] See Klamer and Zuidhof (1998)

3. KEY ELEMENTS IN CULTURE-LED URBAN REGENERATION: LESSONS FROM GLASGOW 1990 AND LILLE 2004

3.1. Glasgow-1990 European City of Culture

The increasing significance of culture-led urban regeneration has urged the city planners to search for culture-led strategies in regeneration plans. Most of them have looked back to the successful or commonly referred practices to derive the key elements for their own city. Since the criteria for success weren't clearly identified, the local authorities were more inclined to search for the practices having long-lasting and sustainable effects. In Garcia (2005), the term of longevity was conceived as a measure for success which the prospects were beneficial inputs for the city and for its inhabitants to survive and develop beyond five years. (Bianchini, 1999; Egan, 2004; Frey, 1999; Urban Task Force, 1999)

Unlike its predecessors, Glasgow was known to be the first ECOC to use the programme as a catalyst for culture-led urban regeneration. The city authorities aimed to extend the buoyant effects as much as possible to guarantee its sustainability. Even though, there have been endless disputes on its diverse impacts -economic, social and cultural-, still the case itself is worthwhile to analyse. It helps understanding why it received wide attention for being a turning point and what could the implications be and then, lead to discussions if it can be named as notorious or famous.

3.1.1 Aim and Objectives

The aim and objectives were documented - including other bidding materials of the programme - in the report *The 1990 story: Glasgow Cultural Capital of Europe* published by the Glasgow City Council in 1992. The principal aim of the ECOC programme for Glasgow was to enhance cultural development by setting up a challenging and an integrative process in hand with public. Other objectives were specified as increasing participation, deepening the conversation of the future image of the city, and increasing media attention. The report does not specifically put forward the targets. As it is conceived, the goals of the Glasgow campaign in the *report* were set out as clear though in a broad manner. Basically, the campaign foresaw to ensure everyone to make the most out of the event. Also, to enhance the city's economic and cultural potential by increasing commercial and cultural investments was stressed among the objectives.

3.1.2 Implementation

After Glasgow was awarded the 1990 European City of Culture, the campaign for promotion launched immediately. The committee consisted of local authorities - city and regional councils- and the tourist board agreed on three target audience groups: Visitors and tourists, opinion-formers and decision-makers and, of course, the people of Glasgow (Glasgow City Council, 1992:23). It was not only aimed to transmit the cultural message to the audience but also to attract the potential business people for unlocking the city's future economy.

Promotion activities were based on four premises, as follows:

Figure-1

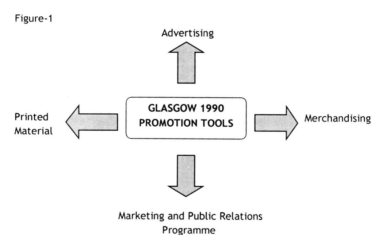

In line with its social goals, many projects had participation and education as goals. Amongst, *Call that singing!*, *Big noise*, *Keeping Glasgow in stitches*, *StreetBiz* (continued after the programme year), *Rarin' to go*, *Glasgow Fair* took place.

Moreover, to stimulate the cultural sector, different cultural fields were encouraged to collaborate for designing cultural programmes and activities and enriching the arts conversation in the city. With the encouragement it was aimed to invest in the cultural infrastructure of the city to boost its potential for development. Programmes fusing both traditional arts and the creative industries played significant roles in transforming the city into a creative hub. Such as the programme "Music in Architecture" was announced as a key driver for the city's architectural heritage. It was set not only mirroring the core architecture of the city but also being instructive in increasing the awareness for the city's architectural potential. In the end, this might also have helped the city to become UK's City of Architecture and design in 1999. This cultural legacy is one of the positive aspects.

The preparation and implementation demanded intense collaborations among diverse actors. The 1990 programme committee was composed by 60 members not only from the Glaswegian artistic and cultural organisations but also from national Scottish institutions, the European Commission and the Office of Arts and Libraries. Besides, public-private partnerships extensively took place and this type of collaboration was usually typified for Glasgow.

For reaching the target audience, Greater Glasgow Tourist Board was in close relations with travel agents and tour operators. The board was also working with the British Tourist Authority's offices abroad to promote the campaign through special "Glasgow 1990" desks. Many telephone kiosks were transformed into information boots to correspond visitors' enquiries.

Since implementation required many actors working together from different fields-government, private, voluntary sectors-, the programme budget necessitated the same diversity. Contrary to the fact that the ECOC programme was introduced by the European Commission, the main financier of the programme was the local government. Only less than 1 per cent of the budget was funded by the European Commission (ibid, p. 30). In the following pie chart, the composition of the budget is depicted.

Figure-2

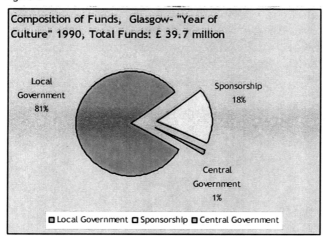

Composition of Funds, Glasgow- "Year of Culture" 1990, Total Funds: £ 39.7 million

Local Government 81%

Sponsorship 18%

Central Government 1%

□ Local Government □ Sponsorship □ Central Government

3.1.3 Outcomes

Little literature was available, besides the *report* published by the Glasgow City Council on specifying the targets. Instead, vague definitions which were counted as overarching goals were mostly embedded in the documents, especially rather in the political ones. Nonetheless, before going in-depth analysis of outcomes under the categories of social, economic and cultural; it is reasonable to provide some concrete figures and statistics which were evidenced by the *report* in 1992. More importantly, it should be better kept in mind that the date the outcomes were publicly accessible it was difficult to make definite conclusions and comments on the intended long-term effects. Therefore, at this point, other academic sources- articles, journals- which were pointing the mid-1990s and onwards were mostly taken into account. For instance, Garcia (2005) has carried out a research on the long-term legacies of Glasgow's ECOC programme and established a model for assessing the cultural impact, which she mainly asserted as the most prevailing outcome resulted from the programme.

Highlights from Glasgow-1990:
- 700 cultural organisations joined from Glasgow
- Nearly 22.000 people were working for organising events
- 1990 projects took place

- All the schools in the Strathclyde region was involved
- 3.439 public events took place
- 23 countries were participating with their performers and artists
- 656 Theatrical productions were realised
- Glasgow City Council provided £ 15 million Cultural Fund to finance 1990 events. In addition to the city council the Strathclyde Regional Council gave away £ 12 million to the 1990 activities which were specialised on education and social work
- 3.5 million visitors came to Glasgow that year

These statistics were compiled from the *report*. However, some are debatable. Especially, in the cultural realm, as it is different from other economic spheres, these statistics and figures is nothing but to demonstrate the quantifiable facts which actually do not reflect the appropriate impact[19]. For instance, above, regarding the last figure on the number of visitors, the *report* hasn't given any extra information whether those visitors all came to attend the ECOC year events specifically or in general to visit Glasgow that year.

Therefore, to proceed to analyse the social, economic and cultural effects will be more plausible to comprehend what has really and to what extent came out and then, step into the problematic issues of this Glaswegian case.

Through out the *report*, it has been explicitly understood that there has been wide media coverage, and in that sense the programme committee has achieved in creating a tumult out of this event. Magnifying media attention was sought as an important tool for driving people's attention to the specific year and thus, to the city's future. After all, the ECOC programme developers wanted the attendance to increase to suffice the future supply of cultural activities which would help transform the city's depressive image from a post-industrialist city to an attractive city fuelling creativity.

A multi-faceted approach for the fulfilment of the year of culture was a necessity for all sorts of benefactors. As it couldn't be possible without the supply-driven factors, the demand side was equally important for the sake of the process. Reaching to the audience did need additional resources due to social inclusion

---

[19] See Pignataro (2003) for interpreting quantitative indicators in the cultural sector.

issues. To address different groups in the community by bringing social work and education programmes was sought as an important initiative for integration. In 1990, Glasgow's year of culture established an education programme aiming long term benefits through innovative projects and activities that would enhance the community's capability and awareness for the event and in return, serve for a better involvement (ibid, 1992).

As a matter of fact, at the end of the year, there was some achievement recorded in increasing the attendance thanks to the first-time attendees. In the museums it was 2 per cent and in dance activities it accounted for 19 per cent of the public (Glasgow City Council Report, 1992:28). In addition, local attendance also increased. Ethnic community involvement was attained. Most of the ethnic communities were involved in organising their local events, activities. Like the Irish came together to perform their traditional folk songs and seize their culture, whereas the Ghanians formed a Steel band and performed to public. Besides, the Jewish groups organised the Festival of Jewish Culture to cherish both the well-known artists in literature, classical arts and music (ibid, 1992). In total, 40 per cent increment in attending theatres, museums and galleries was tracked according to the *report*.

To point out the economic outcomes, the contribution of the ECOC Glasgow programme to the regional economy was £ 7.1-11.1 million in net terms, in 1990. Although the largest part of the financial provision came from the local government, private sponsorship took the second place. Not market development was explicitly stated out, the growth in cultural industries (art trade, music industry, designer trades, film and video, etc) since 1986 accounted for 3.9 per cent by 1991 (Glasgow City Council Report, 1992:30).

Another economic contribution was made to city's cultural tourism and to its business capacity in leisure activities. After the ECOC programme, Glasgow achieved to be known as a chief destination for cultural tourism and a centre for conference and conventions (Garcia 2005; Palmer/Rae 2004).
From the study *European Cities and Capitals of Culture* commissioned by the European Commission[20], it was concluded that specifically the film and music industry stood out to receive more attention in terms of sustainable economic

---

[20] See Palmer/Rae (2004)

growth. Certain development programmes supported these industries (e.g. Glasgow Film Office, the Film City Business Centre, the Music Business Development Programme, the Cultural Enterprise Office)[21].

Among other tangible outcomes, new constructions took place and new iconic buildings rose in the city's cultural appearance. New theatres, parks and refurbished public spaces were counted as novelties during and after the programme. A new department in the city council was also formed to complement new types of collaborations as they appeared to be important for effectiveness and efficiency of the city administration. The city council also, for the first time, began working on a city-wide arts policy, an arts strategy and a cultural industries strategy (Palmer/Rae, 2004:168). Besides, private-public partnership started to play role in the cultural terrain as well. International linkages and networks were expanded during implementation of the programme which in return yielded mutual understanding of different cultures.

All the (positive) outcomes have been discussed so far by citing various sources. Nevertheless, since impact studies of these kinds of programmes were mainly directed to their economic and social outcomes, even they were partly able to be evaluated or measured, there is a new dimension evoking within this discussion. It is the cultural impact seen as a promising field to put more light on (Garcia 2005:1, Evans & Shaw 2004:6).

What the notion of *cultural impact* suggests has mainly to do with image and so forth the national pride. It can either be the impact of a cultural activity on the identity or the image of a particular place, or it can evaluate how the cultural activity affects cultural diversity, citizenship, representation, cultural rights, participation which altogether build up "cultural governance" (Evans & Shaw 2004, Garcia 2005, Mooney 2004).

In her study on Glasgow, Garcia (2005) claims that the most outstanding feature of Glasgow was its cultural impact rather than its economic and social impact. The perception of living in a city like Glasgow, the local and also partially the national pride, the identity of the city has changed drastically within the last decade.

---

[21] Ibid, p. 168

The Glasgow City Council has also denoted in its *report* that many surveys undertaken after the 1990-ECOC programme have revealed that Glasgow was no more attributed as a rusty city but rather as a city increasing its potential in becoming an enjoyable and vibrant place of which people felt more confident and pride.

What an increase of national pride and a positive transformation in the image can bring is more attractiveness to potential business investors. It is hard to say that there can be a direct linkage between the latter and a better city image. However, it is likely that positive city image can influence – indirectly- or be a catalyser in attracting potential business investors. Above all, only concrete measures for the economy can directly pull potential business investors into the city economy.

3.2. Lille-2004 European Capital of Culture

Following the Glasgow case, Lille was another cultural capital which sought the ECOC initiative to function as a steam engine for its culture-led urban regeneration. The process had already initiated in the 1990s by a series of events and decisions resulted in the creation of a (re)development dynamic in Metropolitan Lille and the Region surrounding it (L'Agence de développement et d'urbanisme de Lille Métropole, 2002:23). Similar with Glasgow, Lille was also struggling with its post-Fordist structure which resulted from being formerly a city dominated by heavy industrial sectors-textile, steel and mining- that depressed the city texture and the living conditions. The city is populated mostly with young people having a multi-cultural background counting up to 1.2 million in the Lille Metropolitan region. Alike in Glasgow, Lille already initiated a regional development plan covered through cultural regeneration projects before it was awarded the ECOC title.

Before analysing the case likewise done for Glasgow, the fact that little research undertaken specifically on Lille compared to Glasgow has to be kept in mind. There are indeed several reasons lying behind. First of all, Lille is a very recent example. Compared to Glasgow the city has not accrued its long-term effects yet and thus, so far, a thorough assessment would be difficult within these conditions. Yet still, the case offers much to be searched out in the future.

On the other hand, reviewing the studies of Lille 2004 reveals a more balanced and probably a more optimistic picture than it is for Glasgow. As Glasgow was under harsh attack in terms of discussing social and economic outcomes, it is noticed that Lille has not evoked that much debate yet. Most likely it is due to the reasons mentioned above. However, in the end, looking at the studies done so far Lille 2004 has indeed done far better than Glasgow in terms of clearly defining its targets with highlighting social inclusion and reaching out to wider public specifically and setting the managerial structure in detail.

The subsequent sections will analyse Lille's ECOC programme in depth and demonstrate the novelties specific to the case.

3.2.1 Aim and Objectives

The mission statement of Lille 2004 basically was to regenerate the city by focusing especially on the inhabitants. It aimed a coherent approach covering not only inhabitants in the Lille metropolitan region but reaching to the whole of the Nord-Pas de Calais region and also parts of its neighbouring country Belgium. In this sense, Lille was the first city to extend the event across a whole Euro-regional territory of 193 towns and cities (Lille Metropole 2004, *A-Z Document,* p. 59).

Already perceived as a promising mission, the vision of the programme was described as follows:

> *"We dreamt of Lille as a spaceship changing the fabric of time, a place where everyone can live at their own pace, cross through exotic parallel worlds, stroll through the new frontiers opened up and already dissolved....a process of metamorphosis with the ability and energy to perpetually remodel the world"(Palmer/Rae Associates, 2004, ECOC report Part-2, p.48)*

In the text, the process of *"metamorphosis"* was the narrative used for all the regeneration and refurbishment which was set as a target before the ECOC programme started. Lille focused on to build the process on a long-term scale. Therefore, the ECOC programme was consisted of a small but an enhancing part of this lengthy project.

As an outstanding feature of the case, social inclusion and community involvement were among the priorities. The programme stressed the importance of social objectives to being integrated into the programme itself. Stated in the second part of *European Cities and Capitals of Culture 1995-2004* report by Palmer/Rae Associates (2004:48), "promoting creativity and cultural experience by bringing art into the streets, organising popular street festivals and encouraging encounters between artists and the public" was the key objective in this context. Additionally, educational and social projects were directed to enhance social cohesion, to encourage attendance to the cultural realm, to reinforce democratic participation and increase awareness and knowledge in migration and cultural diversity.

Above all, setting the target as to become a city sustaining cultural development through improving cultural and non-cultural infrastructure was the core idea in summary.

3.1.2 Implementation

After setting the targets, the City of Lille planned a lengthy list of projects for celebrating the event with aiming a large involvement of both national bodies and wider public. Since the ECOC programme was ambitious and far-reaching in its goals, the scale of the programme offered 2000 projects for the whole year.

The projects, as it was previously mentioned, were based on a range from cultural to educational purposes. Particularly, the context of the cultural programme reflected intensity and integrity together with bringing traditional and contemporary forms of art under the same umbrella. Among some projects that illustrated the innovative image were; *Cars of the Future and Cinema of the Future in Lille, BergArt festival in Bergen, Icons of the XXth Century contemporary music series in Graz, Avignonumériques digital art project.*

Hereby, there is a key implication not to overlook, which is the focus of the programme. The focus is not only on the traditional but also on fusing the contemporary with different forms of art to enhance creativity and innovation. In return, the arts conversation is enriched and new approaches have given an impetus for the arts and cultural world.

As new forms in arts were encouraged in this programme, indeed the historical elements were not disregarded in the programme. Being a Flemish city in history, the City of Lille also designed several projects to cherish its Flemish legacy which was an indispensable part of its identity.

Gaining wide public attention also urged the city to make all the activities available and accessible in public space. This was mostly succeeded by projects calling for large involvement. Such as the project *TEXTO Lille*, searched out how the shared memories and the storytelling were in different cities across the region.

The reason behind seeking to achieve wide public involvement departed from the city's rich multi-cultural textile. Since the city was receiving immigrants from different origin due to its industrial past, Lille had an extra challenge to accomplish social cohesion by designating projects accordingly with this goal. Many artists and showcases from different countries abroad, ranging from Africa to Japan, took place in this event. For instance, *Les Afriques* exhibition was reflecting the contemporary African art.

The overarching mission of the programme entailed diverse operating structures for backing up the programme and ensuring the longevity of the regeneration, or so-called *metamorphosis* process. The goal of crossing over the territories and encompassing cities and towns to reach out for a larger audience seemed not possible to achieve without an effective network built around strong partnerships among various bodies from various fields. Such as Lille, being a novel case amongst other ECOCs for this goal, proved to be unique again for its governing structure of the programme.

Basically, the governing board consisted of 42 members with various backgrounds. Most members were politicians from city and regional authorities, representatives from national authorities, universities, cultural institutions and foundations. In addition, there were also members from municipalities across the region and from Belgium. Within this operating structure one of the outstanding features was the board being divided into three sub-divisions: Cultural, institutional and economic. These operational units were performing in alliance to get the administrative work done. The institutional partners of the Lille 2004 programme included the State and different Ministries, the City of Lille, Lille Metropole, the Nord-Pas de Calais region, the General Council of the department of Nord, the General Council of the department of Pas-de-Calais, the cities of Roubaix, Tourcoing, Villeneuve d'Ascq, Courtrai (BE), Tournai (BE), Mouscron (BE), and the French- speaking Community of Belgium (Palmer/Rae Associate, Part-2, 2004: 346).

Proceeding to the operational team of the programme, where teams are the forefront actors in realising the programme, it can be outlined as follows:
- Programming team
- Administration team
- Technical team
- Communications and public relations (PR) team
- Private partners relations team

It is easily conceivable that Lille has formed the managerial structure in sub-divisions in line with its multi-faceted goals. Within the communications and PR team a novel system was developed as it was one of Lille's unique characteristics in the ECOC programme. A system of "ambassadors" based on volunteerism in which approximately 16000 volunteers from all different professions and ages were working for the programme in the large-Lille metropolitan region. It is therefore important to see how promoting the programme to the broadest as possible and encouraging volunteerism is important for the sake of the communications and PR strategy.

Another key factor in building partnerships is to create an environment that fosters collaboration among diverse partners. The private partner relations team was appointed to accomplish this mission. The team not only encouraged different types of partnerships between the national bodies of cities or regions but also constructed new linkages between these bodies and local private businesses. This had also made Lille being referred as a successful case in setting up effective partnerships with local businesses (See Palmer/Rae 2004, Sacco & Blessi 2006).

What has been so far described in detail about the actors in implementation and how it has set up gives raise to the question "What were the financial provisions envisaged to this programme?" The following figure clearly answers the question.

Figure-3

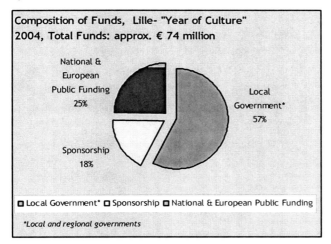

Composition of Funds, Lille- "Year of Culture" 2004, Total Funds: approx. € 74 million

National & European Public Funding 25%

Local Government* 57%

Sponsorship 18%

□ Local Government* □ Sponsorship □ National & European Public Funding

*Local and regional governments

Hereby, when remembering our previous case -Glasgow 1990- it can be observed that both cases are sharing the same trend in devoting local government resources to the ECOC programme compared to other types of funding. Nevertheless, the proportion of national and European funding is slightly higher in Lille's case than in Glasgow's.

Last but not least, the reinforcement of the non-cultural infrastructure amounted over 70 million Euros. This capital investment was mainly funded by the State, the local authorities across the territory and the European Union (through the European Regional Development Fund and the Interregional programmes) (Palmer/Rae, Part-2, 2004, p. 348).

3.2.3 Outcomes

Lille 2004 project revealed both the diversity of the city and drove the *metamorphoses* across the borders. In many ways, the programme helped the city "bringing back its colour". Presumably, the highlights of event, the long-lasting memories and eventually, the hype it created were all repeatedly covered in the media (Sacco & Blessi 2006, Lille Metropole 2005).

Coming to the programme's results, the main highlights were documented as follows:

- Total number of visitors reached over 9 millions (the population of Lille Metropolitan was 1.2 million)
- The opening night event attracted over 650,000 people, which was much higher than the expectations of 150,000 people (Palmer/Rae, Paart-2, 2004: 347)
- 2,500 events took place in 193 cities, towns and villages of the entire Nord/Pas-de- Calais region and Belgium
- Around 39 per cent of the events had free admission in favour of wide participation
- Many blockbuster events took place: the most remarkable was the Rubens exhibition reaching 600,000 attendees in Lille
- 17.000 Artists both from different ethnicities and all around the world took part in this programme
- As a result of the education programmes, more than 1200 schools were involved in the programme
- Visitor attendance to cultural activities increased by 39 per cent
- Partnering with local businesses worked out as increasing corporate sponsoring by 13 million euros, which is said to be the highest ever in the ECOC programme (Sacco & Blessi 2006:12)

Being a recent event and not having much constructive literature does not facilitate the analysis much. So far, what can be put forward is that in overall, Lille produced its own model especially serving as a more socially-friendly ECOC in contrary to what have Glasgow received in respect. It is well-known that those cities becoming cultural capitals of Europe have constructed their own models naturally melted in their specific social, economic and cultural dynamics. Though, learning from the past fellow cities but striving to contribute as another unique model always made the programmes far-reaching sometimes too ambitious. Like Glasgow, even it desired to have a long-lasting economic impact after its ECOC programme the non-measurability and infeasibility limited its potential. Whereas Lille took a more balanced approach by keeping the art programming as the core of the ECOC event. In consequence, in short-term most of the goals were met and not much disappointment was caused among the programme developers.

Starting from the social outcomes, being as the highly respectable field within the literature, Lille has achieved its goal of gaining wide public attention together with increased public involvement. Putting the visitor numbers above mentioned aside, people in regions with no museums got the chance to see travelling exhibitions with national collections of art works. Moreover, around 900 special events targeted youngsters. As Sacco & Blessi (2006) pointed out, the socially disadvantaged areas were priory fields for amelioration with respect to managing social criticalities.

The main concern of community involvement triggered new ways of thinking and enhancing a cooperative network consisted of corporations and cultural institutions together. Indeed, there is clear evidence to show that Lille created its own uniqueness by developing local entrepreneurships (Sacco & Blessi, 2006:13) and diverse types of collaborations among diverse bodies which in return local companies responded positively by letting more than 350 businesses (ibid, p.12) to assist through creating the label "Commerçant Lille 2004". This is what could make Lille 2004 a model considering its operational and managerial structure.

The striking character of the programme was indeed the social projects that were in favour of social cohesion. It can be actually observed that through out history one of the principle objectives of the French cultural policy was to democratise the cultural practices and thus, increase the democratic participation as it was explicitly outlined among the objectives of the ECOC programme. In other words, cultural governance was noticeably at the forefront in all of the activities taken place during the year. The efforts were appreciated for the education and the enhancement of the local community particularly targeting schooling ages and organising workshops among the local and international cultural producers which enriched the mutual understanding of the arts world. A key theme called "metamorphoses" was introduced to the public to challenge their perceptions of the cities and to make them look at the urban transformation from different spectacles. The main actors of this event were artists, designers and sculptors. They transformed the urban environment with extraordinary installations. Apart from this, there was another institution that came to the fore with this programme that is to say, the project called "Maisons Folie". A renovation of 12 historical buildings-factories or sites of the local heritage- across the region and neighbouring Belgium turned into arts centres where around 300,000 people from the local public engaged interactively with artists. Also in the literature, Maisons Folie was cited as a remarkable cultural legacy of the ECOC programme.

As it was reviewed in Palmer/Rae Associates' study report on the assessment of all European Cities and Capitals of Culture, it has been witnessed that most of them have started to expand their focuses towards embracing the economic benefits besides the social and cultural benefits of the programme. Eventually, that raised the bar of expectations in terms of economic outcomes. However, the evaluation reports published or commissioned by each city's own committee right after the event included the changes in economic facts by either stating quantifiable data or assessing the economic development not without relating to its measurability.

Coming to Lille's case, what have assessed under this category so far has not matched the expectations yet. One possible explanation is that the event's outcomes are still continuing and the time span has not arrived to the point in which long-term legacies and effects can be straightforwardly assessed. Hereby, in the meantime while analysing the economic outcomes of Lille 2004 programme; it will be more down-to-earth to speak in terms of changes in infrastructural issues that later on would contribute to the sustainability of cultural development of what was aimed. In addition, other developments and changes traced in different sectors are sketched out within this analysis.

In accordance with the urban regeneration vision, many investments took place to rehabilitate both the cultural and non-cultural infrastructure. Diverse plans were designed in heritage renovation; urban planning and public facilities were implemented by the state and local authorities. Besides the 12 Maisons Folie heritage renovation project, as mentioned above, many other historical sites of the local patrimony were restored. Such as the Hospice Comtess, being transformed from a hospital into a museum; the Opera House of Lille went under complete renovation, whereas Parc de la Deûle and the Tri Postal were other projects which one was a rehabilitation of a park to display land art and the other, an urban transformation of a former three-storey building for postal workers into an iconic venue respectively.

From a sectoral viewpoint, there have been some developments or at least enhancements in the cultural industries. The dynamic environment that helps nurturing innovation and creativity in all forms of arts was in conjunction with Lille's vision. Through out the programme, creative industries were stimulated and kept as an important part of the city's development. Many projects devoted to young talents in design and architecture received wide attention. Moreover, the emerging sectors within the creative industries such as video games and digital arts were also widely supported through several projects called *Game On,* an interactive exhibition tracing the history of electronic games and *Avignonumériques* which was an experimental laboratory for investigating new cultural forms with modern technology.

These activities were indeed important in giving a stimulus for new sectors to flourish. Especially, preserving innovation and creativity was seen essential for the pace of Lille's urban (re)development (L'Agence de développement et d'urbanisme de Lille Métropole, 2002:24). The innovative vision is in line with developments happening in the creative world. As more creativity is cited as the core subject of cultural development which serves for fostering the creative industries, Lille proves to be a wholehearted committer.

Besides the cultural and creative industries were promoted with the ECOC event, another sector, the tourism business also recorded increments in terms of job and business growth. Starting from numbers of visitors, the tourist office has tracked an average of 66% increase between December 2003 and November 2004.
In addition to the average increase of foreign visitors reaching up to 50%, the Dutch customers received a record with a 160.3% rise whilst the average increase of customers in hotels raised by 24.6 %.

In Figure-4, it is depicted that all the complementary sectors to the tourism sector have grown.

Figure-4

Job Growth, October 2003-September 2004

	National Level	Région Nord-Pas-de-Calais	Lille
Sector of Commerce, Hotel, Food & Drink Services, Culture	1.1%	1.5%	7%
- Cultural activities	0.4%	4%	22%
- Retail	1.2%	1.1%	3%
- Hotel business	0.5%	Stable	15%
- Food services	1.4%	3%	7% (13% in the city centre)

Source: Insee-Association Lille Horizon 2004, Impact de Lille 2004 sur l'économie du tourisme in Indicateurs de Lille 2004, p.8

Lille 2004 programme has proven to be a figurehead in concrete efforts for spreading cultural supply by taking social issues as principal. Even it might be still early to provide definite conclusions on the impact of the Lille 2004 programme, what can be clearly said is that Lille's international profile has improved.

4. LESSONS FROM GLASGOW 1990 AND LILLE 2004: PROBLEMS AND CHALLENGES

Glasgow and Lille were two European cultural capitals exemplifying how the ECOC initiative influenced cities as being a mean for their urban regeneration process of what helped rebuilding the cities' image and identity. Although having different motives behind, the fore idea was to stir up public attention for rethinking cities' future development.

Both cities designated as ECOC set the bar high though without specifying the programme objectives in detail. However, it was clear that they had a common belief for the ECOC that should serve as turning point, especially for setting culture-led strategies for urban regeneration. When the ECOC programme ended, both cities ended up with a better outlook thanks to urban renewal projects, having more tourism revenues, an improved international profile to be known as cultural destinations and more attracted visitors and investors. Nevertheless, what was more at stake were the critics being highly posed to the issues that were swept under the rug within all this process.

Indeed, besides all the positive or glittering effects of perceiving ECOC as a thrust to culture-led urban regeneration, this chapter transcends the boosterism part and endeavours to address the problematic and challenging parts of instrumentalising culture in urban regeneration within the ECOC context.

4.1. Economic Problems

This category probably is lying mostly at the heart of the debate: an economically-motivated vision in the ECOC initiative which collides with the aim of pure cultural engagement. As it is stated manifold, priorities have gone through changes where the contribution of the ECOC programme with its rich cultural envisions has been stressed for future economic development of cities. Such an attempt has fallen short and instead, given rise to other counteractive effects in economics that prove not to be healthy and conclusive.

Firstly, measurability of success of the ECOC programmes generally stands out to be the most questionable issue. Cities after celebrating the ECOC event convey statistics of visitors to indicate them as an important measure of success. Despite the fact that visitor numbers are usually used for indicating increments in attendance to events, analysing these indicators should be handled with care especially, when the quantifiable indicators are dealing with the cultural sector. For instance, although the number of visitors in Glasgow reached a peak of 3.5 million within the event year, 1990, it is tricky to say that this figure can be completely attributable to the ECOC event. Moreover, given the rise in attendance rates for both cases, it is inconclusive if there was new audience development or not, which is important for fulfilling a genuine social engagement.

Secondly, as Jones and Wilks-Heeg (2004) have pointed out, culture should not be seen as the sole key driving tool for economic development. There is no clear evidence of 1990 being a direct catalyst for other successes in terms of Glasgow's economic development such as the consistent growth in inward investment and job creation (Palmer/Rae, Part-1, 2004 p. 168). Yet still, it can serve as an indirect catalyst in attracting inward investment and refurbishing the cities' images. But beyond, there are several arguments underlining the possible threats that might occur if culture is sought as the primary instrument for achieving sustainable economic development. These threats are two-sided as it is both equally harmful for the economy and culture.

From an economic viewpoint, flagship cultural events do give an impulse to places where they are celebrated but they do not bring long-lasting effects unless backed up with further concrete measures. Such as, if economic measures for supporting the growth of cultural industries and more correctly, creative industries are taken, then in return it would be more likely to sustain the economic development. Specific measure are needed because those industries hold different capacities to boost creativity and innovation that leads to higher-value added products improved by advanced technology.

Also doubts were posed at the employment opportunities that were repeatedly claimed as an economic outcome of large-scale flagship cultural events. Most of the time, culture and tourism are the highlighted leisure sectors for generating new jobs. Nevertheless, many studies have diluted this claim by pointing out that low-paid service sector jobs are mainly produced from these events (Wynne 1992, Mooney 2004, Jones & Wilks-Heeg 2004). On the other hand, Sacco and Blessi (2006) stirred the discourse by asserting that culture and tourism were high-value added sectors in the economy whereas Bakucz (2007) notes that these sectors do not require high capital formation per job.

From both cases, growth in leisure industries-hospitality, tourism, retail- has been documented and analysed in the study at hand. As Mooney (2004:8) puts forward, in Glasgow during the preparation for 1990 and onwards, significant growth in employment in the 'hospitality', tourism, retail and leisure sectors were noteworthy as sources of more poorly paid, casualised and 'irregular' forms of work. However, Mooney has also stated that Glasgow has not kept up the pace in employment growth recently (2004:9). Moreover, low employment and high unemployment levels have been recorded, though, it is not clear enough to derive that the recent unemployment incident totally a legacy from the ECOC event or to say it is the pure outcome of the culture-led urban regeneration.

The third disputable issue is the attempts to remake the image of cities in order to attract more investors and tourists. Jones and Wilks-Heeg (2004) have acknowledged that these attempts helped transformed the formerly depressed industrial centres both physically and in appearance. In contrary, there were studies voicing the fallacy of *image gilding* in attracting inward investment in reality and asserting the counteractive effects in employment creation (Gomez 1998a).

Together with these image remaking tools too much emphasis on tourists brings fears as well. Both in Glasgow and Lille, the campaign was also positioned on increasing the tourism potential. When both programmes ended, tourist numbers were repeatedly mentioned through out the official evaluation reports. What is more of concern is the inclination of thinking of tourists over the inhabitants. Such an approach was evidenced in Glasgow. The benefits of massive public expenditure were limited to international tourists and the socio-economically privileged communities (Gibson & Stevenson, 2004:2). There is a widely accepted knowledge that initiatives like ECOC increase tourism and convention revenues, whilst little knowledge prevails if this increment can transcend the short-term. More importantly, it is argued that tourism development is uneven across regions and localities (Bakucz 2007:4).

4.2. Social and Cultural Problems

City renewal projects backed up with large scale flagship cultural events especially under the name of 'genuine social engagement' calls the need to focus the analysis on the realised effects in the broader public and understand whether the objectives other than social have posed any potential social and cultural problems or not.

Initially, the arguments were positioned on the accelerating patterns of inequality and thus, gentrification pertained to urban regeneration (Jones & Wilks-Heeg, 2004:2). It was also stressed out that regeneration could sometimes result in tensions caused by sharply rising property values, access to jobs, and competing interests and objectives (Keen et. al, 1989:52). In Glasgow, it is said that a "dual city" character has become more salient than ever. Due to its physical renewal alongside the city's large peripheral housing estates, this has been considered as residual backwaters of dependency, poverty and crime (Mooney & Danson, 1997; Danson & Mooney, 1998). From another arguable viewpoint placed for Glasgow by Damer (1992) was that a refurbishment favouring yuppification could damage the re-industrialisation and impoverish the middle class.

In the plans of Glasgow 1990 and Lille 2004, encouraging social inclusion through integrating more social work projects was set among the priorities. This is indeed quite understandable that the ECOC programme cannot be thought without the public support. Otherwise, neither the city dwellers nor the city itself will benefit from the programme.

Another significant matter arises in this discourse of social inclusion. As the issue of fulfilling social cohesion became so dear in both two cities' ECOC project, and even beyond, frequently cited in cultural development plans, one has to question the sincerity of these projects. It is because to comprehend if the social vector of the project is really the chief part or just a supplement for the disguised aim of economic accumulation (commercial development). As it is evidenced in Damer's quotable words for Glasgow 1990:

> "..All decisions about urban development were hidden from the people and
> so that, the democratic process went out of the window. (Damer, 1992:8)"

However, even Lille has staged a better performance in enhancing democratisation in participation rather than what Glasgow did, the measurability of success is limited within the numbers of attendees and the numbers of social projects. What is more appealing in these two cities' project is the lack of clarity in defining a specific plan to develop new audience in the cultural realm. Sadly, the increase in attendance numbers does not tell us about the composition of the attendees. It would not be inaccurate to say that there is a certain profile of attendees. Especially in a multi-cultural based society, the ethnic minorities are generally the socially deprived ones who are sometimes unintentionally out of the target audience of the cultural events. It would have been much healthier to know whether, let's say, there were minority groups attending or not. It is well known that new audience development is vital not only to engage socially disadvantaged groups into the conversation but also for maintaining social justice in urban development. Then, the "genuine social engagement" would be achieved.

When a city is awarded the ECOC title and declares its main motives to enrich and celebrate the city's culture by aiming high participation rates, hereby, a key question pops up: "Whose culture is celebrated?". A clear-cut answer does not exist when the cities' multi-cultural tissue is taken into account and it can be even harder to find an answer for cities having many inhabitants of diverse ethnic background. There are critics pointing out the unfair situation that the event is accorded to a small group whose benefactors mostly being elitists or investors who will accrue capital thanks to the prospective returns earned from the event. More sadly, some say tension between these cultures may emerge and even cause "culture wars" between the elitists and the so-called 'others' (Jones and Wilks-Heeg 2004, Mooney 2004, Mitchell 2000). So, that is to say "who owns" and "what is represented" are crucial to think upon, like it has always been a matter of possession to Glasgow that "Glasgow ECOC had more to do with selling Glasgow as a place for inward investment than as a celebration of Glasgow culture and Glaswegian life" (Mooney, 2004:5).

When analysing the problematic issues from the economic viewpoint, we denoted that using arts and culture firstly for economic purposes and other mixed-used development projects might not meet the expectations and more importantly, it can also be detrimental for the sake of the cultural content. The risk at stake for Bianchini (1990) is that as the ECOC title provides a wide spectrum of arts activities and incorporates the arts into urban "growth coalitions", this reduces artistic freedom and weakens the core value of arts (in Garcia 2004a). Besides, it increases commercial pressure on arts and culture, in turn which also may be detrimental (Keens et. al 1989). Another problematic issue appears, as the place-marketing project launches for the ECOC event. It implies that culture will be packaged as "official culture" which in return constrains the diverse and challenging cultural activities (Jones & Wilks-Heeg, 2004:13).

4.3. Other Problematic Issues

The ECOC programme necessitates not only well-defined objectives but also a well-developed implementation plan to facilitate the management process and to illustrate which strategies (i.e. communications, partnership etc.) will be followed during the operation of the programme.

Within this hectic procedure many roles are designated to diverse actors having different backgrounds. Mainly, three-parties are involved in the implementation: Public sector (national governmental bodies, local governmental bodies at provincial and municipal level), private sector (companies, sponsors etc.) and the voluntary sector (mostly non-profit institutions, non-governmental institutions-NGOs- cultural organisations, universities, schools etc.). They all share the same goal, which is to do their best to succeed. However, by nature, clash of interests among the parties do happen.

Even it was hard to find evidence from Glasgow 1990 and Lille 2004, it would not be an overshooting prediction if we say that there are certain times where not all parties easily collaborate. A common problem which is quite often encountered during building partnerships is bringing the business people and the cultural producers to the same table can be difficult and turn into a failure sometimes. These two parties have different types of organisational structures and thus, have different languages to communicate. But somehow, it does work when the mission and goals are clearly set and it comes to be everybody's concern. Then, both parties agree to partner to fully commit their responsibilities within a common language. However, when things do not work out in partnering, then that is when vast resources are wasted and leads to undesirable outcomes which can easily danger the prevalent socioeconomic structure of the place and thus, cause a loss in profile.

Looking from the political side, another debatable issue is how cultural policy has moved from the margins to the primary focus of the political agenda. A typology is tended to be composed from former ECOC experiences (as it was done after Glasgow 1990) and aimed to be implemented as a route for other forthcoming cultural capitals. The rising trend of embracing cultural policy as a route for urban transformation has been criticised that it can do no more but glossing over and diverting the attention away from the major structural problems (Mooney 2004). Even urban regeneration prospects were sought under the ECOC project, the long-term objectives were not clearly defined in both cases, Glasgow and Lille. Further beyond this, as a noteworthy detail, the popularity of Lille tended to be attributed more to its beautification project of the Palais des Beaux-Arts than to its 2004 ECoC programme (Newman 2005:67).

4.5. Challenges and Future Topics

The main challenges in the ECOC programme for culture-led urban regeneration differs among cities. It depends on the scale of cities, the socio-economic dynamics and the institutional structure which in fact all add to the specific city culture somehow. It is important to set the process right in accordance with these facts. Cities being either in small or large scale have different advantages and disadvantages in their own right. For instance, in the first part of the *European Cities and Capitals of Culture* report, it is found out that large metropolitan cities face issues of visibility of the ECOC event, attendance, finance, the scale of already existing cultural activity and infrastructure (Palmer/Rae, Part-1, 2004:46). On the other side, small scale cities might struggle with the lack of financial resources which are, in contrary more easily attainable for large cities. When Lille 2004 and Glasgow 1990 are compared, due to Lille's extensive ECOC project embracing many towns and its neighbouring country, the financial resources were much more accessible and vast for Lille than it seemed for Glasgow. Apart from this, within the implementation period, building up an effective communication strategy can be another challenge especially when cities become larger in scale. Like in Lille, after a criticism was directed to its communication strategy the programme organisers responded by creating programmes in many different formats, including monthly programmes and leaflets on specific projects (Palmer/Rae, Part-1, 2004:80). Above all, the challenge remains in various issues as it is also prominent in reaching adequate resources.

In previous chapters and also in the current literature, the measurability of the impacts of cultural investments in general and the culture-led urban regeneration process in particular was highly problematic. For instance, for some cultural investments in the UK important indicators of success for the Tate Modern were stated as employment, commercial development and rising property values (McKinsey, 2002 cited in Newman, 2005:64). Specifically for the ECOC programme, many impact studies were undertaken to follow-up the long term effects. Although these studies sought to pursue quantitative assessments, cultural projects were not quite suitable for being measured by a single approach. Their subject matter necessitated interdisciplinary evaluation methods. Even though Garcia (2004a) suggested that ECOC's cultural regenerations are likely to be measured and justified in non-cultural terms, she instead swivelled the focus on the importance of cultural impact rather than the rest. In addition, Newman (2005) also acknowledges the fact that success is mostly associated to the intangible but the crucial factor of city image.

Beyond all, the real challenge appears as the expectations rise while using the ECOC programme as a catalyser for culture-led urban regeneration. As more dimensions are added into the programme's vision, the process itself becomes more complex and turns out to be non-friendly for carrying out all those impact studies. It is therefore not to be overlooked that suggesting the ECOC as a good strategic and operational basis for culture-led urban regeneration might be misleading somehow as it is poor in standards of event monitoring and evaluation, particularly in the long term (Garcia, 2004a). Even the ECOC is meant to play an essential role in culture-led urban regeneration with together sustaining the process in the future; interestingly, from both cases it is deduced that no concrete strategy for the continuation and follow-up of the programme has been developed (Palmer/Rae, Part-2, 2004:351).

5. ISTANBUL- ECOC 2010: "THE CITY OF FOUR ELEMENTS"

This chapter consists of two sections. The first is an introductory section, mainly describing Istanbul's 2010 programme and it serves for a comprehensive analysis before giving prospects. The focus of the second section is on the implications and challenges of the ECOC programme for Istanbul. Furthermore, it gives prospects for Istanbul 2010 in order to make its experience unique and more importantly, underpins the essential points to sustain a longer legacy left beyond the ECOC year.

5.1. Review of the Istanbul- ECOC 2010 Programme

On 13th November 2006, Istanbul was officially designated as the ECOC for 2010 by the European Council in accordance with Decision 1419/1999/EC, together with other cities, Essen (Germany) and Pecs (Hungary). After the official declaration, the Istanbul ECOC 2010 initiative committee launched a four year action plan to determine the projects and activities prior and during the year of culture, 2010. A wide range of projects displays the rich and diverse content of an ambitious mission to principally accomplish a far-reaching urban redevelopment through arts and culture.

The mission has been identified under five sub-titles by the initiative committee:
1. Urban Transformation: To encourage the inhabitants to participate in the art and culture projects
2. Cultural Heritage: Adding value to the city's cultural assets by preserving the city's cultural and industrial heritage through restoration and regeneration projects that will contribute to their sustainability.
3. Cultural and Artistic Infrastructure: Investing in the artistic and cultural platforms to support all types of artistic performance, arts exhibitions, libraries, education and media that in turn will increase community involvement.
4. Multiculturalism: Stimulating all kinds of activities and projects for sustaining the rich diversity Istanbul has remaining from its patrimony.
5. Cooperation Teamwork: Promoting different types of partnering among diverse actors from diverse sectors, including public, private, volunteer sectors (NGOs and non-profit organisations). Not only during

implementation but also in decision-making bodies, these partnerships will be formed.

These five properties of the mission statement are also embedded in the following multi-faceted future vision including: urban transformation, the conservation of historical and cultural assets, protection against earthquakes, transportation, tourism planning, the planning of residential areas, industrial transformation, an evaluation of labour force potential, cultural industries and improvements in the quality of urban life (Çalışlar, 2005, Section II:6).

The objectives set below show that ECOC serve as an important instrument in changing and enhancing many things in Istanbul. Such as;
- Providing an opportunity to initiate projects that address the social dimensions of cultural policy and action
- Building new capacity by redevelopment projects
- Creating a platform of new administrative models that will stir up the discussion among different urban participants and thus, giving the opportunity to look at new forms of civic activism and urban participation
- Setting up a bottom-up approach in the managerial process will also help intensifying the debate and encourage participation
- Starting from 2006 till 2010, partnering with other fellow cultural capitals to develop projects for a better mutual understanding and ensuring them to be active participants of Istanbul ECoC 2010

It is thus apparent that the ECoC programme is not limited to a short-term incident. A long-term project is envisaged on receiving long-lasting and sustainable returns. In the following *Box.1*, the initiative committee have sought the benefits as:

Box. 1

The Benefits for Istanbul

- o From 2006 onwards the name of Istanbul will be associated with culture and the arts all over the world.
- o As Turkey moves ahead with the process of its candidacy for the European Union, the projects that will be realized will demonstrate that Istanbul, the symbol of the country, has been interacting with European culture for hundreds of years.
- o The city's cultural heritage will be managed in a sustainable manner and it will become even more of a magnet than ever.
- o Istanbul will achieve lasting gains in the fields of urban renewal, urban living and environmental and social development.
- o New museums will be established to protect and display our cultural assets and historical buildings will be renovated, given new roles and opened to the public.
- o Istanbullites will embrace new artistic disciplines. Young talented people will have the opportunity to become more closely involved in artistic creativity.
- o Jobs will be created for a large number of people ranging from communications to organization, education, design, management and creative fields.
- o Those who come to Istanbul for cultural and artistic projects will visit the city's cultural riches, mosques, churches, palaces and museums.
- o Cultural tourism will be invigorated and develop. (Educated, cultured tourists spend three times as much as normal tourists. This means that, as a European Capital of Culture, Istanbul will have a great tourism potential.)
- o Many people from the world of culture and the arts, together with members of the print and visual media, will come to Istanbul from Europe and different countries all around the world.
- o This will make a positive contribution to the promotion and branding of Istanbul.
- o Being selected as a European Capital of Culture will give a boost to the city's economic relations with Europe as well as contributing to its cultural relations.
- o With the renovation that will take place, the administrators and administered will join together, hand in hand, sharing their knowledge and experience, to develop a long-term sustainable model for the future.
- o Through discovering the beauty of their city, Istanbullites will be proud to live in such a city.

Source: Istanbul 2010 official website: www.istanbul2010.org

5.1.1. Projects for ISTANBUL 2010

Since Istanbul 2010 launched its campaign with the slogan as "Istanbul: A City of Four Elements", many projects and activities prior and during 2010 are going to take place. The projects will describe why the city is made up of those elements. Beyond, it is aimed to increase cultural supply by enabling inhabitants to have the opportunity in reinforcing the city's image together. These projections will also contribute to build a solid background for future cultural development.

Briefly, the initiative committee has sought a four season programme each connected to an element and mirroring Istanbul's unique characteristics respectively. The main elements symbolising each season are matched as: "Earth-1 January-20 March (Winter): Tradition and Transformation", "Air- 21 March-21 June (Spring): Heaven Sent", "Water-22 June-22 September (Summer): The City and The Sea", and "Fire-23 September-31 December (Autumn): Forging the Future".

The planned projects are all rich in content and innovative in fusing different forms of arts that nurtures a creative environment suitable for cultural and creative industries to flourish.

From the sectoral breakdown of projects[22] which will take place in 2010 the majority, amounting 27 out of a total of 76 projects, will be in performing arts. In particular, music, theatre and other multidisciplinary forms will be shown under performing arts. Apart from these, 20 of them will be cultural events and festivals; 19 projects will be related to visual arts, again with the combination of multidisciplinary forms and design together; and the rest will be including other fields of the creative sectors like audiovisual, architecture and new digital media.

Prior to the event year 2010, 14 projects on urban regeneration, transformation, museums and heritage restoration will be accomplished. More importantly, many other social projects including seminars, symposiums, and social work under specific districts which are going under urban renewal projects will be realised to fire up the local inhabitants.

5.1.2. Organisation and Implementation

A decree by the Turkish cabinet enacted in 2005, legitimized the Istanbul 2010 initiative committee and its duties. According to the decree, a hybrid organisation was set up involving non-governmental organizations, the Municipality of Istanbul, the Governorship of Istanbul, the Ministry of Culture and Tourism and the Ministry of Foreign Affairs, universities and individuals. Indeed, unlikely the predecessors of the ECOC programme, this type of organisation mostly rooted on volunteerism suggested to be a deliberate type of organisation, whereas the alternative could have been filled with professionals. However, the latter was contradictory to the initiative committee's mission in increasing public involvement and thus, justifying a bottom-up approach.

After Istanbul being officially recognised as the ECOC of 2010, two sub-committees- Artistic and Communication- linked to the executive board was established in addition to the advisory board. The former is responsible for implementation,

---

[22] For the sectoral break down of projects and for the full-list of 72 projects see APPENDICES I-II

coordination and evaluation of the arts activities to be held in 2010, while, the latter is in charge of bringing experts from the PR and advertising sectors together under an in-house creative team to design and develop strategic planning documents and implement communications plan respectively.

Likewise in the cases Glasgow 1990 and Lille 2004, reaching to the broadest public and making almost all the inhabitants aware of the cultural capital programme necessitated a comprehensive communications and marketing strategy, which has obviously taken into greater account.

Huge effort is going to be given to ensure that inhabitants are kept informed during the preparations and when the event takes start.

In 2007, the objectives and the target audience of the communications strategy have been set in *Box.II* as follows:

Box.II

Communication Strategy Objectives

- Increasing public participation/attendance
- Enhancing reputation of the institutions/communities
- Meeting the goals and objectives initially determined
- Integrating the PR campaign with the overall communications and marketing plan
- Combining established advertising means with guerrilla marketing and grass-roots communications in order to maximize the collective impact

Target Audience

- The global community
- Europeans
- Global and local universities and institutes
- Global and local NGOs
- Focused study fields: urban sciences, arts and culture, architecture
- Funding agencies
- Policy makers and opinion leaders at local and national levels
- The media
- Industry
- The public

Source: Çalışlar (2005), Section V, p.3

It seems like the expansive target audience has not eased the implementation of the communication strategy. Moreover, vast resources might need to be channelled in order to reach the above mentioned objectives.

In addition to developing strategic planning materials with the PR and advertising experts, the communication committee will make further points clearer to this group that the main targets are stated to raise the international profile of Istanbul, change the image of the city, make Istanbul as a well-known destination for the foreign and domestic tourists, expand the audiences for culture, promote the city and its entire activities related to its title ECOC, develop and promote multicultural policies and stimulate partnerships.

Some of these targets which are specifically stressed in the promotion campaign are though questionable in many senses. This will be resumed when prospects and challenges are explained for Istanbul's case.

In order to achieve the targets, starting from 2008, series of activities and tools of the harsh campaigning have been listed in *Box.III*:

Box.III

Series of communication activities

- Organising a non-media channels system of civic organisations, chambers of commerce & trade, related associations, university campus and community leaders
- Compiling a list of media contacts and figures which includes radio, television, web, and local publications
- Promoting partnerships with industry
- Designing and developing printed materials such as brochures and posters
- Creating a series of news releases and distributing the releases to media channels
- Visiting local radio channels and TV stations to create opportunities for the promotion of the ECoC concept
- Preparing the press releases and on-air announcements
- Organizing periodic news conferences
- Carrying out the publication of a weekly electronic ECoC 2010 Istanbul newsletter
- Developing and periodically updating the ECoC 2010 Istanbul multilingual web site

Tools of the communications strategy

- Initiating creative and alternative media channels and tools such as:
- SMS, cell phone, GPRS technology
- ATM machines
- Public areas (WCs, mosques, stadiums, supermarkets ...)
- Licensed merchandising
- Limited and specially designed POPs
- Special events

Source: Çalışlar (2005), Section V, p. 3

From all the ambitious goals, strategies and diverse projects the Istanbul 2010 ECOC programme has made it clear that immense financial resources are required.

In the following pie chart (Figure-5) the composition of funds is quite peculiar with respect to Glasgow and Lille, in which both of the largest proportions were resulted by the local governments. The reason of the peculiarity lies in the distinct funding scheme of Istanbul 2010. In the official application document of Istanbul 2010 prepared by the initiative committee (Çalışlar, 2005) the total amount of the planned budget is 120.3 million Euros. 84 per cent of the budget, the most bulky part accounting for 99.8 million Euros, is planned to be obtained from a special levy which will be specific to Istanbul. The imposition of this special levy will be put into effect by the Turkish Government through issuing a decree that will charge 1 YKr (one hundredth of 1 YTL (New Turkish Liras), equivalent to US $ 0.6 cents or € 0.4 cents) per litre of gas or oil sold in Istanbul from the date Istanbul is declared the ECOC for 2010. The remaining proportions of the budget are nearly identically shared by sponsors and the central government amounting for 8 per cent each, and with a very little financial support from the European Union (EU) funds. As a matter of fact, very recently, the European Commissioner for the EU Enlargement Mr. Olli Rehn has promised a contribution of 1.5 million Euros for the ECOC 2010 initiative[23]. How the new grant will be distributed among the objectives is another question which has already led to public debate.

Figure-5

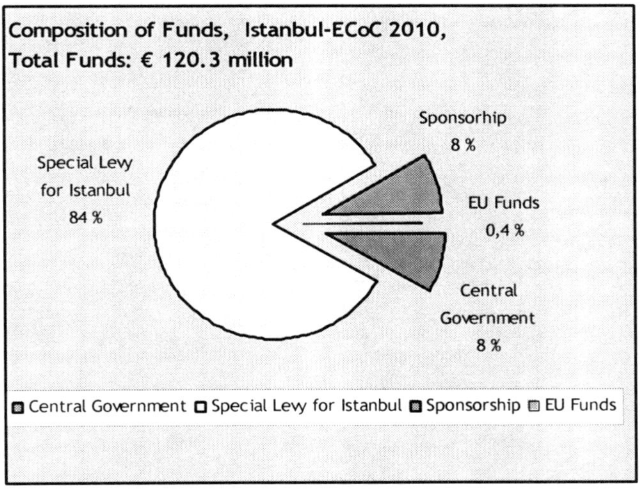

Composition of Funds, Istanbul-ECoC 2010,
Total Funds: € 120.3 million

Special Levy
for Istanbul
84 %

Sponsorhip
8 %

EU Funds
0,4 %

Central
Government
8 %

☐ Central Government ☐ Special Levy for Istanbul ☐ Sponsorship ☐ EU Funds

5.2. Challenges and Prospects

The debate on the implications of the ECOC 2010 programme has taken start. Minimising the probability of undesirable effects necessitates to initially be acquainted with the challenges lying ahead so that the process can be healthily

---

[23] For detailed information see Karahan, J. *Discord threatens Istanbul 2010 European Capital of Culture project,* in Today's Zaman March 9, 2007, website accessed on 25th June 2007: http://www.todayszaman.com/tz-web/detaylar.do?load=detay&link=104868

managed and expectations can be rationally set accordingly. Hereby, it is intended to provide a list of the possible challenges and prospects for Istanbul 2010 in the light of the viewpoints including economic, social, cultural, organisational and operational, which all have facilitated for a comprehensive analysis of the interaction between the cultural regeneration and the ECOC in particular understudied in this present thesis.

- First and foremost, ECOC should be expected to be not a direct but more functioning as an indirect catalyst for raising the international profile of the city. In order to elude a fallacy in economic terms, it would be reasonable not to place too many expectations on cultural industries. Like, the new museums and galleries may not have the economic capacity to engage regeneration and community development. Nevertheless, undoubtedly the cultural impact will result in pride and confidence.

- More job opportunities will eventually be created. However, the fact that these are mostly consisting cyclical or seasonal jobs should be kept in mind. A long-lasting impact can hardly be obtained unless concrete political measures are taken for strengthening the environment for creative and cultural industries. At this point, responsible authorities and policy makers need to reconcile by taking concrete measures. A consent on taking the initiative for stimulating the cultural and creative industries can only be given after all interested groups discuss together.

- In order to evaluate the objectives cautiously, the results can be conveyed into an in-depth research which would help capture the cultural, social and economic impacts during and after the ECOC event. This would also benefit the future process in rethinking the subject of culture-led urban regeneration.

- For an enhanced artistic and cultural development and preserving artistic diversity, it would be better to not incorporate the economic motivation into the arts programming of the ECOC programme.

- A transparent setting needs to be systematically adopted in all mechanisms of the ECOC initiative in order to make it clearer for whom it will be beneficial. In addition to transparency, non-bureaucratic characteristics would also benefit for an effective and efficient operation.

- Understanding the motives of the target audience is indeed important for developing local demand which in turn sustains the supply of new cultural assets (Atkinson, 2005).

- It is quite understandable that the current political practice is attaching social objectives to urban cultural regeneration purposes, yet still, there are tensions that may erupt among the social and economic objectives.

- Regarding the high rate and dispersed structure of suburbia in Istanbul, a holistic cultural planning should not be overlooked. There could be more rigorous social projects set out for community engagement.

- Avoiding elitist approaches will better serve towards a genuine social cohesion.

- It should be kept in mind that policies without effectively focusing on the causes of social problems will be more likely to fail. For instance, through arts and culture it is aimed to increase social cohesion by eradicating social exclusion. However, before shaping such urban cultural policies, it is thus essential to find out the root of the problem and define its causes to treat the symptoms. Not all socially excluded groups are the same. Atkinson (2005) makes a distinction by pointing out that some are excluded largely due to the powerlessness rather than choice and some can be resulted from unfavourable employment circumstances, which happen to be more acute in the rapidly changing Istanbul. Especially, those who are excluded due to long-term unemployment need different emphasis via taking specific labour market measures which also help them to invigorate the social ties with the community. Otherwise, cultural policies in isolation will not be able to heal the social wound.

- Use of multi-sectoral partnerships in the urban regeneration projects is crucial, though it is also important to bear in mind that many of the problems which partnerships have experienced stem from an unwillingness to recognise that they involve power relations (Atkinson, 2005).

- Since the Istanbul 2010 initiative committee has prioritised urban redevelopment through arts and culture with the help of the ECOC programme, there are certain challenges that are faced during the urban regeneration process: joined up approach to urban problems, community participation and addressing (urban) social exclusion (Atkinson, 2005).

- It is observed that the Turkish government has little role in putting their hands on the ECOC project. It is due to a deliberate decision to perform all the activities with a bottom-up approach and in a voluntary team spirit. Even it is not expected from the government to plan all the urban regeneration projects, the government can still proceed in allocating the required resources and performing some tasks, especially in training the project participants' creative skills and managing externalities.

- More corporate network is needed especially for the prospective years especially when strategising for arts and culture and also strengthening the cultural infrastructure. It could be hardly realized without developing a corporate network together with the local authorities.

- Istanbul, considered as one of the large metropolises in Europe, the projected budget itself remains to be a challenge. The special levy implemented for Istanbul will also help indicate how this kind of financial scheme can work out for a metropolitan case.

- According to the recent report (Palmer/Rae, Part-1, 2004:51) published on the past European cultural capitals, the lessons show that objectives of each ECOC should follow from its broad mission, and be compatible, evaluating carefully where the attainment of one objective prevents the attainment of another. Therefore the objectives should be limited in number and prioritised, ambitious but achievable, measurable in some way, kept under review and integrated into the entire process of planning.

- The Istanbul ECOC 2010 will also be a challenge for city governance as well. It can serve as a model for the city's future development if only it succeeds in establishing a strong dialogue between governmental, private and the voluntary sectors.

- Even the application bid came from a bottom-up approach; the authorities at the top should have the capability and sense of perception how the event for the public and the city as whole is really important.

- It is important to clearly set the target audience. Is it the foreigners or are the Istanbullites to celebrate the event? Is it sufficient to attach the motive of reaching to the broadest public when a vast number of target groups are considered from the communication strategy?

- Last but not least, Istanbul is already a potent symbol for its traditional heritage. It has the potential to blend the traditional heritage with contemporary culture in which innovation and creativity is heavily ingrained. The increased use of technology unlocks the innovation capacity which could give birth to many variations of new forms in arts and culture. Enriching the spectrum of arts and culture with creativity will assist cultural industries to move its momentum.

6. CONCLUSION

In the beginning of this present study, the main question in mind was initially "Why cities are pursuing culture-led strategies in urban regeneration?" and then understand "How can culture contribute to urban regeneration in social and economic terms?". The answer was strived to be found through performing a comparative analysis seeking what could be learned from the cases of Glasgow 1990 and Lille 2004 in light of the *European Capital of Culture* (ECOC) initiative. The main motive to ask these questions was to finally provide prospects for Istanbul's 2010 ECOC programme and hence, open the floor to future discussions upon the emerging challenges.

Following the introduction part, the second chapter conceptualised the idea behind the culture-led urban regeneration phenomenon in order to answer the "Why" question by stating factors that engendered this process. The main findings were manifold:

i. In the aftermath of the decline of manufacturing industries that led to declining cities, the early 1980s has manifested that increased competition among cities to secure inward investment, expand their external markets and attract more visitors urged them to re-imagine their international profiles and therefore eventually place cultural promotion tools at the centre of their political agendas.

ii. The social pillar attached to culture was brought up via stimulating community identity, social cohesion and cultural governance.

iii. The adhesive role of culture especially in multicultural societies was promoted as an indispensable element in social development objectives.

iv. The paradigmatic shift in cultural policies and the diminishing role of public intervention encouraged private sector to involve in local cultural programming.

Besides, evidence from the immense literature was stated for the justification of the importance of culture through both looking from the economic and social perspectives.

Before undergoing into an in-depth analysis, background information was also necessary to provide for comprehending the ECOC initiative and the role of Istanbul in this research and also its importance for the ECOC initiative. Not it can be confined in few words, though, Istanbul's competitive and comparative advantages in being a large metropolitan case and the unlike stance among its predecessors as being a historically dominant, polycentric and a non-declining city point out that it is *hopefully* expected that the ECOC experience will bring many challenges and insights in urban governance and social cohesion issues.

Continuing on to the third chapter, after understanding the rise of culture in urban regeneration the second attempt was to investigate "How" culture could contribute to cities' urban regeneration process. Hereby, a comparative analysis was conducted among two case studies deliberately chosen from the well-known cultural capitals, Glasgow 1990 and Lille 2004, which have sought the ECOC initiative as a stimulating tool for urban (re)development purposes. Another important point to note down was that both cities shared common traits as being heavily industrialised in history and then later on becoming declining cities due to their post-Fordist structure.

The case studies were analysed according to their aims and objectives they had for the ECOC programme, the implementation procedure and finally their outcomes. Put it simple, these two examples have demonstrated how they carried out the implementation by showing their peculiar marketing and communications strategies and their projected operational system. What was most memorable from the Glasgow 1990 and Lille 2004 was the successful partnerships in common, the strong image rebuilding and the voluntary system respectively. Above all, from the case studies evidenced in this chapter revealed that culture did not seem to be a 'cherry-on-the-cake' anymore.

Besides underlining the key elements in culture-led urban regeneration driven from the cases above mentioned, it was more of importance to scrutiny the research towards highlighting the problematic issues which appeared to be generally swept under the rug. The fourth chapter dealt with these issues mainly; to also ground the study on a two-sided approach that helped to discuss the most disputable economic, social, organisational and other problematic issues repeatedly cited in the literature. As it was noted previously, such a critical investigation also enlightened Istanbul's case and facilitated to understand the lessons drawn from Glasgow 1990s and Lille 2004.

The main problems in using the ECOC as the main driver for urban regeneration emerged from high economic expectations and the lack of clarity in long term projections. All cases implementing a multi-faceted approach were cherished for their diversity. However, when economic motives were considered it was found out that an economic-driven arts programming might not be sustainable neither for the urban economy nor for the social cohesion. Unless culture-led urban regeneration was backed up by concrete measures directed to economic issues and social issue separately, then the process would rely on a much healthier track which could lead to buoyant effects. Putting more concretely, if more employment opportunities are aimed to be created it should be kept in mind that creating jobs solely through cultural activities is neither effective in terms of permanency and nor it brings high-value adds to the overall economy. Besides using cultural promotion tools, to address the economic problems, if it's an unemployment problem let's say, then, it would have been far better to incorporate labour market measures. This is to say, putting culture on top of the political agenda as a cure for both social and economic problems will be nothing more than a reckless approach that in turn might cause other detrimental effects.

For the last but not the least chapter, beyond stating content and projections the 2010 ECOC programme for Istanbul, it was more insightful and suggestive in dealing with future implications and providing prospects to help realise both the positive and negative attributes of the phenomenon. Since measurability had always been a widely debatable issue during monitoring and evaluating the ECOC process and other outcomes of the cultural events, it would be a wise attempt to think on the tools for assessing Istanbul's forthcoming ECOC experience. This would bring on fruitful discussions and opening more room for future research.

Another topic to be left for further research is dwelling more on the question whether using cultural policy will be effective in solving urban problems of non-cultural nature. An empirical research held on the latter topic would indeed stir the current conversation. Furthermore, it might ground the rhetoric that has been constructed so far by mostly policy makers as seeing culture as "the" solution.

So far, the best words to trace out the challenge for Istanbul's case can be put forward as;

> *"Large cities today may well harbour unprecedented creative capabilities, but they are also places where striking social, cultural, and economic inequalities prevail, and there can be no truly final achievement of the creative city where these stubborn problems remain. This is not simply a question of income distribution, although more equitable economic conditions for all must surely figure prominently on any agenda of reform (Scott, 2006:15)"*

In the end, the last words remain for the argument. Today, the role of culture cannot be underestimated in strengthening social ties within community and for its products that stimulates new forms to fuse across different creative sectors which all are essential components of the new economy or in other words, the creative economy. Within this context, it is always good to revisit the "How and what you build through cultural industries?" question in light of these developments. Specific for the ECOC programme, there are two ways to avoid failure: (i) Either to set the ECOC just as a simple cultural event lasting for a year (ii) Or, if the ECOC is set as an catalyser for creating urban regeneration, then it would be better to ensure that the process is based on all parties trust and commitment to bring on relevant policy responses and reforms to acute problems that could not be solely solved by cultural measures. Otherwise, in the latter, by rising too many expectations in the culture-led urban regeneration process to overcome prevalent problems through ECOC will alas lead to a bitter end, which I do not intend to comment any further.

APPENDICES

APPENDIX I: Sectoral Breakdown of Projects

Sectoral Breakdown of Projects

	<2010	In 2010
Artistic and Monumental Heritage (Urban renewal	<u>14</u>	-
projects)	-	-
- Historical monuments	5	-
- Museums	7	-
- Archaeological sites		
- Other heritage	2	-
Visual Arts	-	<u>19</u>
- Visual arts (incl. design, exhibition)	-	11
- Photography	-	2
- Multidisciplinary	-	6
Performing Arts	-	<u>27</u>
- Music	-	20
- Dance	-	-
- Music Theatre	-	-
- Drama Theatre	-	3
- Multidisciplinary	-	4
- Other (circus, pantomime, etc.)	-	-
	-	<u>7</u>
Audio and Audiovisual Media/Multimedia	-	5
- Film	-	-
- Radio	-	-
- Television	-	-
- Video	-	-
- Audio records	-	-
- Multimedia	-	2
Book and Press	-	<u>1</u>
Architecture	-	<u>2</u>
Archives	-	-
Events (cultural events, walking tours, symposiums etc.)	-	<u>20</u>

APPENDIX II: LIST OF PROJECTS
KEY

PArt: Performing Arts
VArt: Visual Arts
Ex: Exhibition
MD: Multidisciplinary
Event: Cultural event, symposium, festival

Earth > Tradition & Transformation
1. Grand Opening (Music)
2. Mothers, Goddesses And Sultanas (Ex)
3. Imperial Passions (Ex)
4. Istanbul Inspirations (Music)
5. Turkish Makam Music, Between The Past And The Future (Music)
6. Istanbul Costumes (Ex)
7. Topkapi Palace Cyber Museum (VArt-MD)
8. Only In Istanbul (Music)
9. International Puppet Theater Festival (Theatre)
10. 7000 Years At 7000 Meters (Event)
11. Harmony in Diversity (Music)
12. Şenlikname (Event)

Water > The City and The Sea
36. Europe On The Bosphorus (Ex)
37. Painting The Bosphorus Blue (Ex)
38. Istanbul History And Sea Festival (Event)
39. 40 Hammams Story (Ex)
40. The Danube Bridge (PArt-MD)
41. From The Past To The Future: Istanbul And Her Sister European Capitals Of Culture (Event)
42. Istanbul Meeting Of The European Capitals Of Culture (Event)
43. Europ-Ist 2010 (Event)
44. By The Flow Of Water (Event)
45. Youth Integrart (Music)
46. 3 Countries - 3 Composers - 3 Concerts (Music)
47. International Istanbul Music Festival (Music)
48. International Istanbul Jazz Festival (Music)
49. Far Away So Close (Music)
50. Heybeliada Sound Project (Music)
51. International Festival Of Islands (Event)
52. Ramadan Festivities 2010 (Event)
53. Mahya Contest (Event)
54. Children's Festival: Children's World (Event)

Air >Heaven Sent
13. Living Together (Ex)
14. Icons And Sacred Relics In The Atrium Of Hagia Eirene (Ex)
15. Hagia Sophia Throughout The Centuries Symposium (Event)

Fire > Forging The Future
55. International Istanbul Biennial (VArt-MD)
56. Architectural Biennial (Architecture)
57. Istanbul On The Move (Architecture)
58. Design 2010 (Design)

16. Islam & Humanitas (Event)
17. Days Of Faith (Event)
18. Hidrellez (Event)
19. 40 Days 40 Concerts (Music)
20. Babylon Turns Istanbul On (Music)
21. Major Encounters (Music)
22. New Language Of Music In New Europe (Music)
23. Tango: 3 Cities, 3 Cultures And A Passion (Music)
24. Istanbul: A Refugee For All Times (VArt-MD)
25. Towards Heaven (Event)
26. The Meeting Of Street Theatres (Theatre)
27. 2010 District Festivals (Event)
28. International Istanbul Theatre Festival (Theatre)
29. 3 Women / 3 Istanbul - Bodies And The City (PArt-MD)
30. Thinking Together On Public Projects (Event)
31. International Istanbul Film Festival (Film)
32. 10 Istanbul (Film)
33. Istanbul, Symphony Of A City (Film)
34. We're Uncovering Istanbul (Film)
35. The Immigrants - Towards A Common Future (Film)

59. International Student Triennial (Ex)
60. Istanbul: A Place With No Doors Or Windows (VArt-MD)
61. Dem(Art)Racy Village (VArt-MD)
62. Forging The Future, Forging Culture (Event)
63. Photo-Bridge (Photography)
64. International Istanbul Photography Festival (Photography)
65. "4+1" Earth/Air/Water/Fire + Eternity (Design)
66. 20th Akbank International Jazz Festival - Celebrating Istanbul@ The European Crossroads (Music)
67. Mediterranean Contemporary Music Festival (Music)
68. Perafest (VArts-MD)
69. Two Musical Geniuses From East And West: Itri & Bach (Music)
70. Miam Electro-Acoustic & Intermedia Platform (Multimedia)
71. Miam Film Music & Sound Design (Audio-Multimedia)
72. Miam Contemporary Music (PArt-MD)
73. Miam Ethnomusicology (music)
74. Impressionist (PArt-MD)
75. The Conference Of World Artists For Peace (Event)
76. In Lieu: Writing On/In/To Istanbul (Book)

REFERENCES

Atkinson, R. 2005. Urban Regeneration, Partnerships and Community Participation: Lessons from England. In Kucukcekmece Municipality Publication. *Istanbul 2004 International urban regeneration symposium: Workshop of Kucukcekmece District.* Istanbul, 119-129.

Bakucz, M. 2007. European Capital of Culture on the Periphery. Conference paper in *Regional Studies Association International Conference, Regions in Focus.* 2-5th April 2007, Lisbon. Website: http://www.regional-studies-assoc.ac.uk/events/pastevents.asp

Bianchini, F. 1990. Urban renaissance? The arts and the urban regeneration process, in: S. MACGREGOR and B. PIMLOTT (Eds) *Tackling the Inner Cities: The 1980s Reviewed, Prospects for the 1990s*, pp. 215-250. Oxford: Clarendon Press.

Bianchini, F. 1993. Remaking European Cities: The Role Of Cultural Policies, In: F. Bianchini et al. (eds). *Cultural Policy and Urban Regeneration: The West European Experience.* Manchester: Manchester University Press, 1-21.

Bianchini, F. & Parkinson, M. (eds) 1993. *Cultural Policy And Urban Regeneration: The West European Experience.* Manchester: Manchester University Press.

Bianchini, F. (1999) Cultural planning for urban sustainability, in: L. Nyström and C. Fudge (Eds) *Culture and Cities: Cultural Processes and Urban Sustainability*, pp. 34-51. Stockholm: The Swedish Urban Development Council.

Çalışlar, İ. (eds.) 2005. *Candidate for 2010 European Capital of Culture, a City of the Four Elements.* Istanbul: Mas Matbaacılık.

Crane, D. 1992. *The Production of Culture: Media and Urban Arts.* Newbury Park, CA: Sage

Damer, S. 1992. *Glasgow, European City of Culture 1990: Images and Realities.* Amsterdam: Centrum voor Grootstedelijk Onderzoek.

Danson, M. & Mooney, G. 1998. Glasgow: A Tale of Two Cities: Disadvantage and Exclusion on the European Periphery, in Lawless, P., Martin, R. and Hardy, S. (eds.) *Unemployment and Social Exclusion*, London: Harper Collins

Egan, 2004. *Skills for sustainable communities.* Report to the Office of the Deputy Prime Minister. London: RIBA Enterprises Ltd.

Evans, G. 2001. *Cultural Planning: An Urban Renaissance?* London: Routledge.

Evans, G. 2005. Measure to Measure: Evaluating the Evidence of Culture's Contribution to Regeneration. *Urban Studies* 42 (5/6)

Evans, G. & Shaw, P. 2004. *The Contribution of Culture to Regeneration In the UK:*

A Review of Evidence: A Report to the Department for Culture, Media and
Sports. London Metropolitan University: London.

Florida, R. 2002. *The Rise Of The Creative Class.* New York: Basic Books.

Frey, H.1999. *Designing the City: Towards a More Sustainable Urban Form.*
London:E & FN Spon.

Garcia, B. 2004a. Urban Regeneration, Arts Programming And Major Events.
International Journal of Cultural Policy., 10(1), 103.

Garcia, B. 2004b. Cultural Policy and Urban Regeneration in Western European
Cities: Lessons from Experience, Prospects for the Future. *Local Economy*
19(4), 312-326.

Garcia, B. 2005. Deconstructing the city of culture: The long-term cultural legacies
of Glasgow 1990. *Urban Studies* 42(5/6)

Gibson, L. & Stevenson, D. 2004. Urban Space and The Uses Of Culture.
International Journal of Cultural Policy 10(1), 1

Glancey, J. 29 March 2003. Bright Lights, Big City. *The Guardian.*
http://arts.guardian.co.uk/cityofculture2008/story/0,,950353,00.html

Glasgow City Council. 1992. *The 1990 story: Glasgow Cultural Capital of Europe.*
Department of Performing Arts & Venues. Glasgow: M&M Press

Go´mez, M.V. 1998. Reflective images: the case of urban regeneration in Glasgow
and Bilbao. *International Journal of Urban and Regional Research* 22(1),
106-21.

Gratton, C. and Taylor, P. 1995. Impacts of Festival Events: a Case Study of
Edinburgh. In: Ashworth, G. et al. *Tourism and Spatial Transformations.*
Wallingford, CAB International: 225-238; CCPR, Glasgow University
(www.gla.ac.uk)

Hitters, E. & van Aalst, I. 2005. The place 2B: exploring the logic of urban cultural
clusters. Submitted to *International Journal of Urban and Regional
Research.*

Hitters, E. & Richards, G. 2002. The Creation and Management Of Cultural
Clusters, *Creativity and Innovation Management,* 11 (4), 234-247.

Hitters, E. 2007. Porto and Rotterdam as European Capitals of Culture: Toward
Festivalisation of Urban Cultural Policy. In: G. Richards (eds.). *Cultural
Tourism: Global and Local Perspectives.* Atlas: The Haworth Press. 281-301

Isar, Y.J. & Bonet, L. 2007. *Mapping the Cultural Economy in the Euro-
Mediterranean Region.* Mediterranean Programme 8th Mediterranean
Research Meeting, Florence and Montecatini Terme, 25-27 March 2007.

European University Institute- Robert Schuman Centre for Advanced Studies. Consulted in 26 March 2007
http://www.iue.it/RSCAS/Research/Mediterranean/mrm2007/pdf/WS9_MR M2007.pdf

Jones, P. & Wilks-Heeg, S. 2004. Capitalising culture: Liverpool 2008. *Local Economy,* 19 (4) 341-360

Karahan, J. 9 March 2007. Discord threatens Istanbul 2010 European Capital of Culture project. *Today's Zaman,* website accessed on 25th June 2007: http://www.todayszaman.com/tz-web/detaylar.do?load=detay&link=104868

Keens, W., Owens, P., Salvadori, D. And Williams, J. (eds.) 1989. *Arts and the changing city: An agenda for urban regeneration.* New York: British American Arts Association

Kelly, A. & Kelly, M. 2000. *Impact and Values -Assessing the Arts and Creative Industries in the South West.* London: Bristol Cultural Development Partnership

Klamer, A. and Zuidhof, P.W. 1998. *The role of the third sphere in the world of arts.* Unpublished paper. Erasmus University Rotterdam/Faculty of History and Arts

Klamer, A. 1996. *The Value of Culture.* Amsterdam University Press: Amsterdam

Kunzmann, K. R. 2004. *Culture, Creativity and Spatial Planning.* Abercrombie Lecture. Liverpool University/Department of Civic Design

Landry, C. 2000. *The Creative City; A Toolkit For Urban Innovators.* London: Comedia/Earthscan.

Lavanga, M. 2006. *The Contribution of Cultural and Creative Industries to a More Sustainable Urban Development: The Case Studies of Rotterdam and Tampere.* Draft paper for the ACEI (Association of Cultural Economics International) Conference Vienna, 6-9 July 2006

Media Group (Turku School of Economics) and MKW Wirtschaftsforschung GmbH 2006. *The Economy of Culture in Europe.* European Commission. http://ec.europa.eu/culture/eac/sources_info/studies/economy_en.html

Miles, M. 1997. *Arts, Space and the City.* London: Routledge

Miles, S. & Paddison, R. 2005. Introduction: The Rise and Rise of Culture-led Urban Regeneration. *Urban Studies* 42 (5/6), 833-839.

Miles, M. 2005. Interruptions: Testing the Rhetoric of Culturally Led Urban Development. *Urban Studies* 42 (5/6), 889-911.

Mitchell, D. 2000. *Cultural Geography: A Critical Introduction*. Oxford: Basil Blackwell

Mooney, G. & Danson, M. 1997. Beyond "culture city": Glasgow as a dual city, in: N. Jewson and S. Macgregor (Eds) Transforming Cities: Contested Governance and New Spatial Divisions, pp. 73–86. London: Routledge.

Mooney, G. 2004. Cultural Policy as Urban Transformation? Critical Reflections on Glasgow, European City of Culture 1990. *Local Economy* 19(4), 327-340.

Montgomery, J. 2003. Cultural quarters as mechanisms for urban regeneration. Part 1: conceptualising cultural quarters, *Planning, practice & Research*, 18 (4), 293-306.

Myerscough, J. 1988. *Economic Importance of the Arts in Glasgow*. London : Policy Study Institute.

Newman, P. 2005. Cultural regeneration, tourists and city government. In Kucukcekmece Municipality Publication (Eds). *Istanbul 2004 International urban regeneration symposium: Workshop of Kucukcekmece District*. Istanbul, 63-69.

Palmer/Rae Associates. 2004. European Cities and Capitals of Culture 1995-2004 (Part I). Brussels: European Commission
http://www.palmer-rae.com/culturalcapitals.htm

Palmer/Rae Associates. 2004. European Cities and Capitals of Culture 1995-2004 (Part II). Brussels: European Commission
http://www.palmer-rae.com/culturalcapitals.htm

Pine II J. & Gilmore, L. 1999. *The Experience Economy*: McGraw-Hill Ryerson College Division.

Pignataro, G. 2003. Performance Indicators. In R. Towse (eds.). *A Handbook of Cultural Economics,* Edward Elgar: UK, 366-373.

Prentice, R. & Andersen, V. 2003. Festival as Creative Destination. *Annals of Tourism Research* 30, 1: 7-30.

Sacco, P.L. and Blessi, G. T. 2006. *European culture capitals and local development strategies: comparing the Genoa 2004 and Lille 2004 cases*. Vancouver: Centre of expertise on culture and communities, Creative City Network of Canada

Scott, A.J. 1988. *New Industrial Spaces*. London: Pion.

Scott, A. J. 2000. *The Cultural Economy of Cities*. London: Sage Publications

Scott, A. J. 2006. Creative Cities: Conceptual Issues and Policy Questions. *Journal of Urban Affairs,* 28 (1), 1-17.

Sharp, J. & Pollock, V. & Paddison, R. 2005. Just Art For A Just City: Public Art And Social Inclusion in Urban Regeneration. *Urban Studies* 42 (5/6), 1001-1023.

Stevenson, D. 2004. 'Civic gold' rush: cultural planning and the politics of the Third Way, *International Journal of Cultural Policy*, 10, pp. 119-131.

Storper, M., & Christopherson, S. 1987. Flexible specialization and regional industrial agglomerations: the case of the US motion-picture industry. *Annals of the Association of American Geographers*, 77, p.260

Throsby, D. 2001. *Economics and Culture*. Cambridge: Cambridge Press

Tornqvist, G. 1983. Creativity and the Renewal of Regional Life, in Buttimer, A. (eds.) *Creativity and Context: A Seminar Report*. Lund: Gleerup, 91-112

Towse, R. (eds) *A Handbook of Cultural Economics*, Edward Elgar: UK

Urbact Culture Network. 2006. Culture & Urban Regeneration: The Role of Cultural activities and Creative Industries in the Regeneration of European Cities. Conclusions & Recommendations. *Urbact Publications*

Urban Task Force. 1999. *Towards an urban renaissance*. London: E & FN Spon, HMSO.

UNCHS (HABITAT) 2004. *The State of the World's Cities 2004/2005: Globalization and Urban Culture*. Nairobi: UNCHS; and London: Earthscan

Wilks-Heeg, S. & North, P. 2004. Cultural Policy and Urban Regeneration: A Special Edition of Local Economy. *Local Economy* 19(4), 305-31.

Wynne, D. 1992. *The Culture Industry: The Arts in Urban Regeneration*. Aldershot, UK: Avebury

Wu, C.T. 2002. *Privatising Culture: Corporate Art Intervention since the 1980s*. London: Verso

Yardımcı, S. 2005. *Kentsel Değişim ve Festivalizm: Küreselleşen Istanbul'da Bienal*. Istanbul: Iletişim

Zukin, S. 1995. *The Cultures of Cities*. Cambridge: Blackwell

Websites

- Eurostat Online Database, Urban Audit: http://epp.eurostat.ec.europa.eu/ accessed on 28 March 2007

- http://portal.unesco.org/culture/en/ev.php-URL_ID=24478&URL_DO=DO_TOPIC&URL_SECTION=201.html accessed on 25 March 2007.

-
http://www.culture.gov.uk/www.culture.gov.uk/Templates/Publishing/Research.
aspx?NRMODE=Published&NRNODEGUID=%7bC0E53EED-B0BD-41ED-BC1D-
B717723781A1%7d&NRORIGINALURL=%2fReference_library%2fResearch%2fdet%2fglo
ssary_abbreviations%2ehtm&NRCACHEHINT=NoModifyGuest#top accessed on 25
March 2007.

- http://capital.culture.info/ accessed on 26 February 2007

CPSIA information can be obtained at www.ICGtesting.com
Printed in the USA
LVOW08s0959310813

350423LV00002B/468/P

9 783847 306719